GEOLOGISTS' ASSOC]

THE YORKSHIRE COAST

Second Edition

P. F. RAWSON and J. K. WRIGHT

With contributions by :–

J. E. Hemingway

F. Whitham

Edited by J. T. Greensmith

© THE GEOLOGISTS' ASSOCIATION
1992

Notes. *The details of routes given in these guides do not imply a right of way. The onus of obtaining permission to use footpaths and to examine exposures rests with the user of the Guide who should observe carefully the Code for Geological Field Work issued from the Librarian, The Geologists' Association, c/o Department of Geological Sciences, University College London, Gower Street, London WC1E 6BT.*

In particular, those in charge of parties should ensure that there is no indiscriminate hammering of, or collecting from, exposures and that no damage is caused to property.

Any information (e.g. change in footpaths, filling in of quarries, threat to SSI's, new exposures) that would update and improve a revised edition of this guide would be welcomed by the Association.

CONTENTS

		PAGE
Preface		7
Introduction		8

ITINERARIES

I	Staithes to Port Mulgrave P. F. Rawson	23
II	Whitby to Saltwick Bay J. E. Hemingway, revised by J. K. Wright	31
III	Robin Hood's Bay J. K. Wright	37
IV	Blea Wyke Point and Ravenscar J. E. Hemingway, revised by J. K. Wright	40
V	Cloughton Wyke to Scalby Ness J. K. Wright	50
VI	Egton Bridge and Goathland J. E. Hemingway	61
VII	South Bay, Scarborough, Cayton Bay and Gristhorpe Bay J. K. Wright	66
VIII	Filey Brigg J. K. Wright	78
IX	Castle Hill, Scarborough, and the Hackness Hills J. K. Wright	82
X	Reighton Gap to Speeton Cliffs P. F. Rawson	88
XI	Thornwick Bay and North Landing, Flamborough P. F. Rawson and F. Whitham	94
XII	Flamborough Head P. F. Rawson and F. Whitham	100
XIII	South Landing to Sewerby F. Whitham	103
XIV	Langtoft, Foxholes and Staxton Hill P. F. Rawson	109
	References	112

LIST OF FIGURES

		PAGE
Figure 1.	Geological map with itineraries.	8
Figure 2	Structural framework of the region.	10
Figure 3	Structural inversion of the Cleveland Basin.	11
Figure 4	Palaeogeographical setting.	14
Figure 5	Devensian geography.	22
Figure 6	Staithes — Rosedale Wyke shore map.	24
Figure 7	Cleveland Ironstone Formation at Staithes.	27
Figure 8	Cleveland Ironstone Formation and 'Striped Bed'.	29
Figure 9	The Scaur, Whitby, map.	32
Figure 10	Cliff section at Whitby.	33
Figure 11	Geological map of Robin Hood's Bay.	38
Figure 12	Localities at Ravenscar.	41
Figure 13	Section across the Peak Fault at Ravenscar	43
Figure 14	Robin Hood's Bay from the Peak.	44
Figure 15	Localities between Cloughton and Scalby.	51
Figure 16	*Equisetites* bed at Cloughton Wyke.	53
Figure 17	Saurian footprint, Long Nab, and channel sands, Scalby Bay.	57
Figure 18	Section of lower Long Nab Member.	58
Figure 19	Meander belt channels, Scalby Bay.	59
Figure 20	Glacial and pre-glacial drainage in the Egton Bridge and Goathland area.	63
Figure 21	Localities in the Scarborough-Cayton Bay area.	67
Figure 22	Moor Grit and Long Nab Members, South Bay.	69
Figure 23	High Red Cliff, Cayton Bay, and Scarborough Formation, Gristhorpe Bay.	75
Figure 24	Cliff section at Red Cliff, Cayton Bay.	76
Figure 25	Map of Filey Carr Naze and Brigg.	79

Figure 26	Corallian section at Filey Brigg.	80
Figure 27	Corallian, north of Filey Brigg and *Thalassinoides* burrows.	81
Figure 28	Localities in Scarborough and Hackness Hills.	84
Figure 29	Map of the Speeton section.	89
Figure 30	Remanié horizon and ammonites in the Speeton Clay.	92
Figure 31	Localities of the Flamborough area.	95
Figure 32	Chalk sequence at Thornwick Bay and North Landing.	96
Figure 33	Little Thornwick Bay and Thornwick Nab.	99
Figure 34	"Shatter belt" in Selwicks Bay	101
Figure 35	Chalk sequence from High Stacks to Sewerby Steps.	104
Figure 36	Buried cliff at Sewerby.	108
Figure 37	Glaciology of the eastern end of the Vale of Pickering.	111

LIST OF TABLES

Table 1	Subdivision of the Lower Jurassic.	13
Table 2	Subdivision of the Middle Jurassic.	17
Table 3	Subdivision of the Middle to Upper Jurassic.	19
Table 4	Subdivision of the Cretaceous.	21

PREFACE

Since the first edition of this guide was published almost thirty years ago (Hemingway, Wilson & Wright, 1963; revised edition 1968) Britain's offshore oil industry has developed rapidly. In turn, the Yorkshire coast has become recognised as a crucial region for comparison with the geology offshore. This has stimulated much new research, especially on the structural setting and depositional environments of the Mesozoic sediments, so that a revised guide to the area has become necessary. While much of the text is completely new, we are very grateful to Professor J. E. Hemingway for allowing us to modify and update some of his contributions to the original guide.

The increasing school and public interest in geology over the same period has seen a huge increase in the number of parties and individuals visiting the sections so splendidly exposed along the coast. This has led to overcollecting (especially of fossils) from many of the best-known sites, and created pressure on conservation bodies and landowners. Much of the coast is now managed by the National Trust or Heritage Coast, who have created pathways along and down the cliffs: please do not stray from the marked routes. It is rarely necessary to collect fossils *in situ*; excellent specimens can be picked up loose from patches of shingle at many of the localities described.

Before following the coastal itineraries consult the local tide tables and unless an access or escape route is immediately adjacent, **never start work on a rising tide.** The cleanest exposures are often at the cliff foot where there is an ever-present danger of falling rock; take sensible precautions by avoiding areas of recent cliff fall and by wearing safety helmets. Please try to avoid disturbing nesting seabirds.

Appropriate Ordnance Survey (OS) and Geological Survey (GS) maps are listed at the beginning of each itinerary. The location maps are simplified from 1:25,000 O.S. maps. The whole area is covered topographically in a single sheet by Sheet 36 of Bartholomew's National Map Series at the 1:100,000 scale. The Institute of Geological Sciences' Tyne-Tees Sheet (1:250,000) also embraces the whole region.

We thank D. N. Wright and C. R. Ivens for assistance with fieldwork and photographs, and Janet Baker and Colin Stuart (University College London) for drawing the figures.

INTRODUCTION

The Yorkshire coast provides some magnificent exposures of Jurassic and Cretaceous rocks that were deposited in the Cleveland Basin and on the adjacent northern margin of the East Midlands Shelf (Figures 1 & 2). The coastal area is now firmly established as a standard for comparison with both the less well exposed areas inland and also for the offshore North Sea basins.

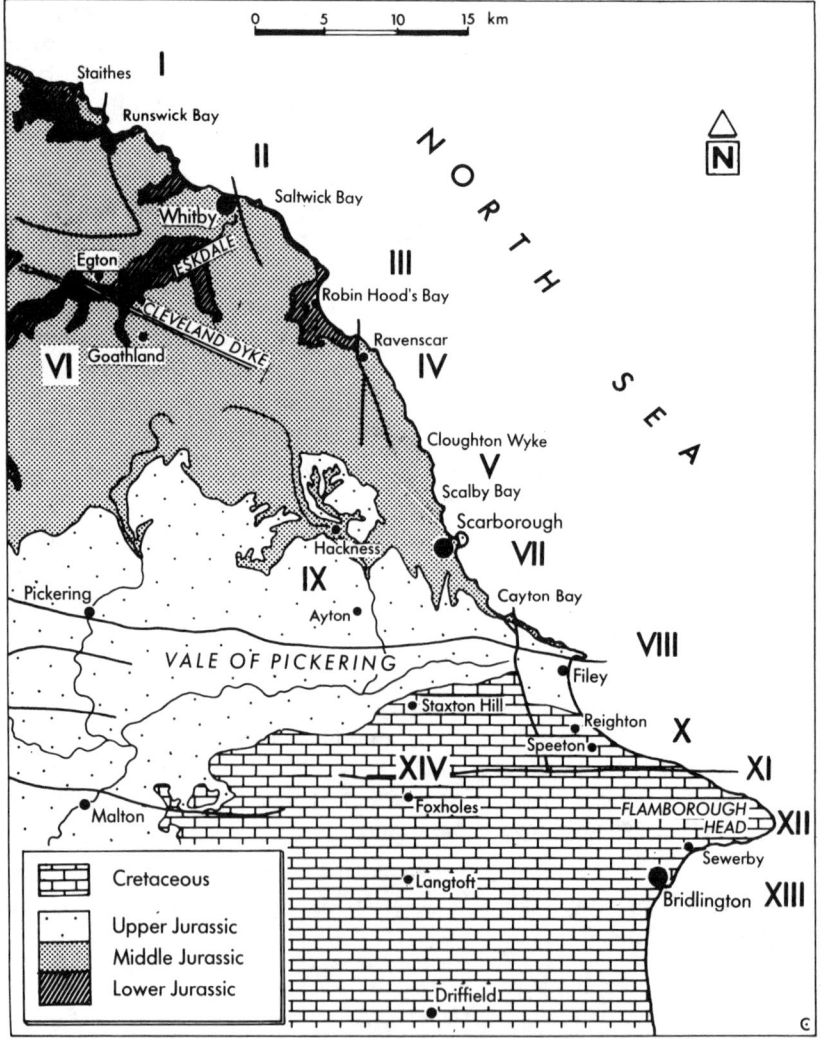

Figure 1. Geological map of the coastal region, with itineraries indicated.

It has attracted the attention of geologists from the earliest days of our science and continues to do so. William Smith visited the area several times in the first two decades of the nineteenth century and recognised almost all the groups of strata that he had previously defined in southern England. Largely through his nephew John Phillips, and particularly after he had settled near Scarborough in 1828, Smith encouraged the publication of descriptions of the fascinating series of rocks he had found. However, local workers were also becoming involved. Young and Bird's *A Geological Survey of the Yorkshire Coast* appeared in 1822. A revised edition was issued in 1828, followed shortly afterwards by John Phillips' *Illustrations of the Geology of Yorkshire; Part 1. - the Yorkshire Coast* (1829). These pioneering works laid a firm though sometimes conflicting foundation for later researchers to build upon.

The structural framework
Towards the end of Triassic (Rhaetian) times the Cleveland Basin and East Midlands Shelf began to develop through differential subsidence, which continued through much of the Jurassic and Cretaceous. Their Mesozoic configuration (Figure 2) may reflect the buried Carboniferous structure (Kent 1980b). The Howardian-Flamborough Fault Belt formed the southern margin of the basin, while its western limit was probably defined by the Pennine High. The northern limit may lie in the Tyne area, as Mesozoic rocks thin in that direction offshore (see maps in Kent, 1980b).

Few faults are known within the basin, except along the coast where several N-S trending ones occur (Figure 2). The best-known is the Peak Fault, whose origin has led to considerable discussion (Itinerary IV). Seismic information offshore now shows that it forms the western boundary to a narrow N-S trending graben, the Peak Trough, which runs obliquely to the coast with the Red Cliff fault forming its eastern margin (Milsom & Rawson, 1989). Fault movement in the trough probably occurred intermittently from the Triassic through to the Tertiary.

As in so many other parts of northwest Europe, basin inversion started in eastern England during late Cretaceous or Tertiary times, when the Cleveland Basin was uplifted to form the present day east-west trending Cleveland Anticline. The axes of several subsidiary domes and troughs are aligned obliquely to the main axis (Figure 3). The Lockton and Eskdale domes have yielded gas, the latter in commercial quantities. The most spectacularly exposed of these minor fold structures is the Robin Hood's Bay dome on the coast southeast of Whitby (Itineraries III & IV; Figure 14).

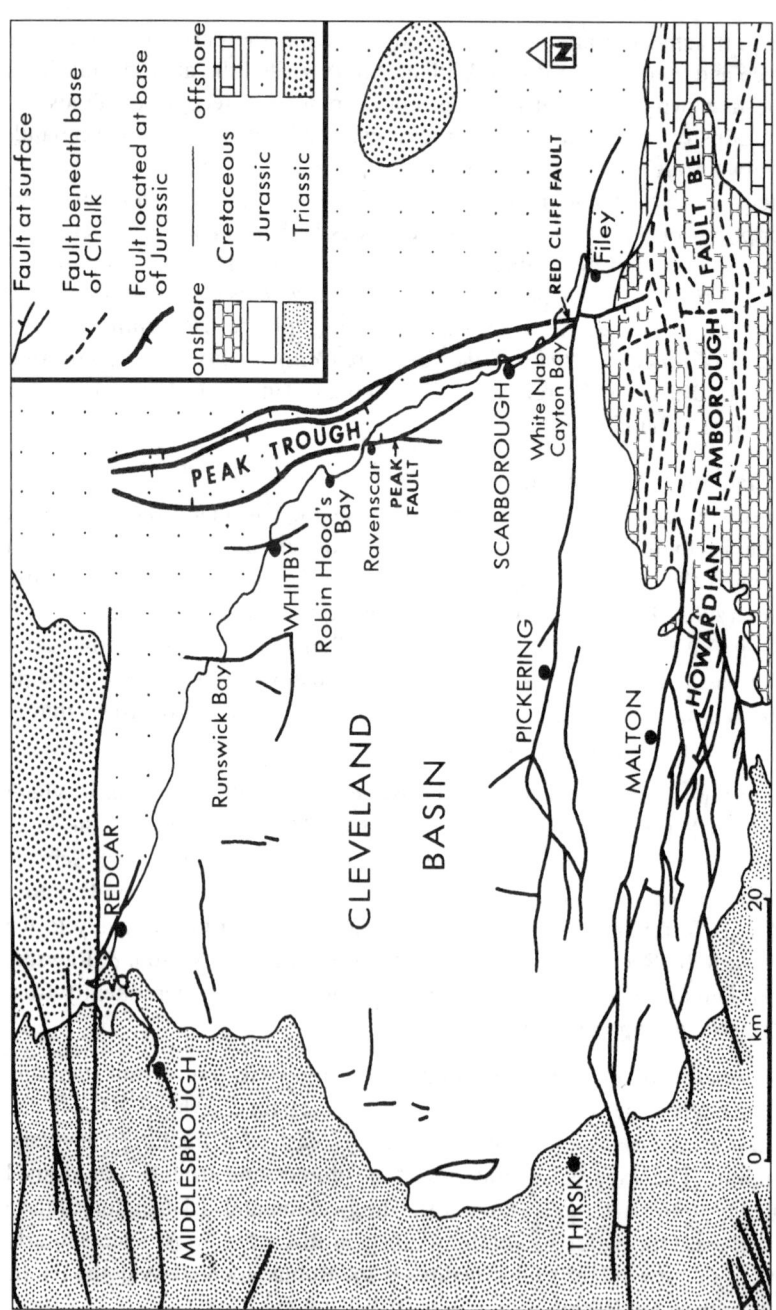

Figure 2. The structural framework (after Kirby & Swallow, 1987, and Milsom & Rawson, 1989)

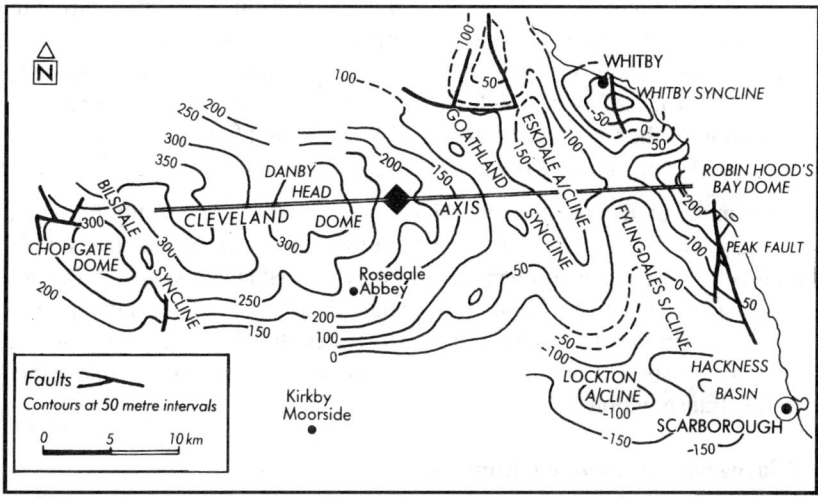

Figure 3. Structural inversion of the Cleveland Basin: as indicated by contours drawn on the top surface of the Dogger Formation (modified from Kent, 1980b, fig. 24, and Hemingway & Riddler, 1982, fig. 4).

The main phase of inversion is assigned either to the late Cretaceous and early Tertiary (Kent, 1980a) or to the Tertiary alone (Hemingway & Riddler, 1982). The latter authors deduced that some 1 to 1.25 km of late Jurassic to Cretaceous and 1.0 to 1.25 km of Tertiary sediments have been stripped off subsequently. While this inversion probably reflects compression from the south, Tertiary igneous activity to the north-west of Britain also impinged on the area, for the injection of the Cleveland Dyke marks a peripheral event in the volcanic history of the Hebridean igneous province.

The northern part of the East Midlands Shelf forms the Market Weighton High. This regionally important structure was originally described as an anticline, but is now regarded as a rigid E-W orientated unfolded block which remained buoyant throughout Jurassic and Cretaceous times (probably supported by a deeply buried granite) while the Cleveland Basin to the north was subsiding rapidly. The high is essentially a hinge between the shelf and the basin, the main line of inflection being the Howardian-Flamborough Fault Belt. The history of faulting in this belt is still poorly known but probably started during the Carboniferous, when it might have been contiguous with the Craven Fault Belt some 140 km to the west. The Jurassic rocks of the Howardian Hills are intensively disturbed by a series of E-W faults, which pass eastwards beneath the Chalk (Kirby & Swallow, 1987). They probably originated as Jurassic growth faults and were intermittently active during the late Jurassic and early Cretaceous (late Cimmerian movements). Thus, at times, a submarine or subaerial fault scarp probably formed the northern boundary of the Market Weighton High. Further reactivation occurred after deposition of the Chalk, indicated by E-W trending faults and "shatter belts" in the Chalk (Itineraries XII and XIV).

For much of Jurassic time the Market Weighton High probably formed an area of shallow water deposition, though those post-Liassic sediments which originally extended over it were removed by pre-Albian erosion (see Tables I-IV for stage/age terms). From mid-Volgian to early Albian times it appears to have been emergent, forming a barrier between the Cleveland Basin and the shelf to the south, until it was finally submerged again by the mid-Albian marine transgression.

Palaeogeography and environments
Sea-level rise in latest Triassic to earliest Jurassic times established a marine regime over the region, which was marginal to the Southern North Sea Basin (Figure 4a). Lower Jurassic shorelines generally lay some distance from present outcrops and the sequence consists predominantly of offshore argillaceous sediments, though shallow-water facies of varied provenance also occur. The Lower Jurassic sediments are placed in the Lias Group, the nomenclature of which has been revised by Knox (1984), Powell (1984) and Howard (1985); five formations are recognised (Table I), while many of the long-established smaller, essentially lithological, subdivisions have been retained and formalised as members.

The Siliceous Shales are the lowest unit to be seen in detail in the itineraries to be described in this guide. They consist of silty shales with thin harder siltstones, the latter forming the harder ledges picked out by the sea in the intertidal zone at Robin Hood's Bay (Itinerary III). Sellwood (1970) showed that each siltstone band formed the top of a small-scale (1-4 m) coarsening-upward cycle, and that this long interval of cyclic sedimentation was terminated abruptly by the basal Pliensbachian sea-level rise which restored deeper water shale sedimentation to the basin.

Later Pliensbachian times saw a further shallowing of the sea and argillaceous sands and silts were deposited to form the Staithes Formation (Itinerary I). Much of the sediment was derived from a Pennine landmass to the northwest. A shoreline in that direction is also indicated by the overlying Cleveland Formation. This consists of a nearshore sequence in Cleveland dominated by oolitic ironstones, passing to the southeast into a more

STAGE / SUBSTAGE	AMMONITE ZONE	LITHOSTRATIGRAPHICAL DIVISION		
182 Ma	Dumortieria levesquei	Blea Wyke Sandstone Formation	Yellow Sandstone Mbr	9 m
			Grey Sandstone Mbr	9 m
TOARCIAN	Grammoceras thouarsense	Whitby Mudstone Formation	Fox Cliff Siltstone Mbr	11 m
			Peak Mudstone Mbr	13 m
	Haugia variabilis		Alum Shale Mbr	37 m
	Hildoceras bifrons			
	Harpoceras falciferum		Mulgrave Shale Mbr	31 m
	Dactylioceras tenuicostatum		Grey Shale Mbr	14 m
UPPER PLIENSBACHIAN (DOMERIAN)	Pleuroceras spinatum	Cleveland Ironstone Formation	Kettleness Mbr	10 m
	Amaltheus margaritatus		Penny Nab Mbr	19 m
	Prodactylioceras davoei	Staithes Sandstone Formation		25 m
LOWER PLIENSBACHIAN (CARIXIAN)	Tragophylloceras ibex	Redcar Mudstone Formation	"Ironstone Shales"	57 m
	Uptonia jamesoni		Pyritous Shales	26 m
	Echioceras raricostatum			
UPPER SINEMURIAN	Oxynoticeras oxynotum		"Siliceous Shales"	40 m
	Asteroceras obtusum			
	Caenesites turneri		"Calcareous Shales"	127 m
LOWER SINEMURIAN	Arnioceras semicostatum			
	Arietites bucklandi			
	Schlotheimia angulata			
HETTANGIAN	Alsatites liasicus			
204 Ma	Psiloceras planorbis			

Table 1. Subdivision of the Lower Jurassic (Hettangian – Toarcian) sequence. The age is indicated in millions of years (Ma).

NOTE: The Mulgrave Shale Member is proposed here as a replacement name for the Jet Rock Member of Powell (1984) (formerly Jet Rock Formation or Jet Rock Series). This is to avoid confusion with the "Jet Rock sensu stricto", the lowest of three informal divisions commonly used within the member (the other two being the Bituminous Shales and the Ovatum Band). The name is derived from Powell's reference section at Port Mulgrave.

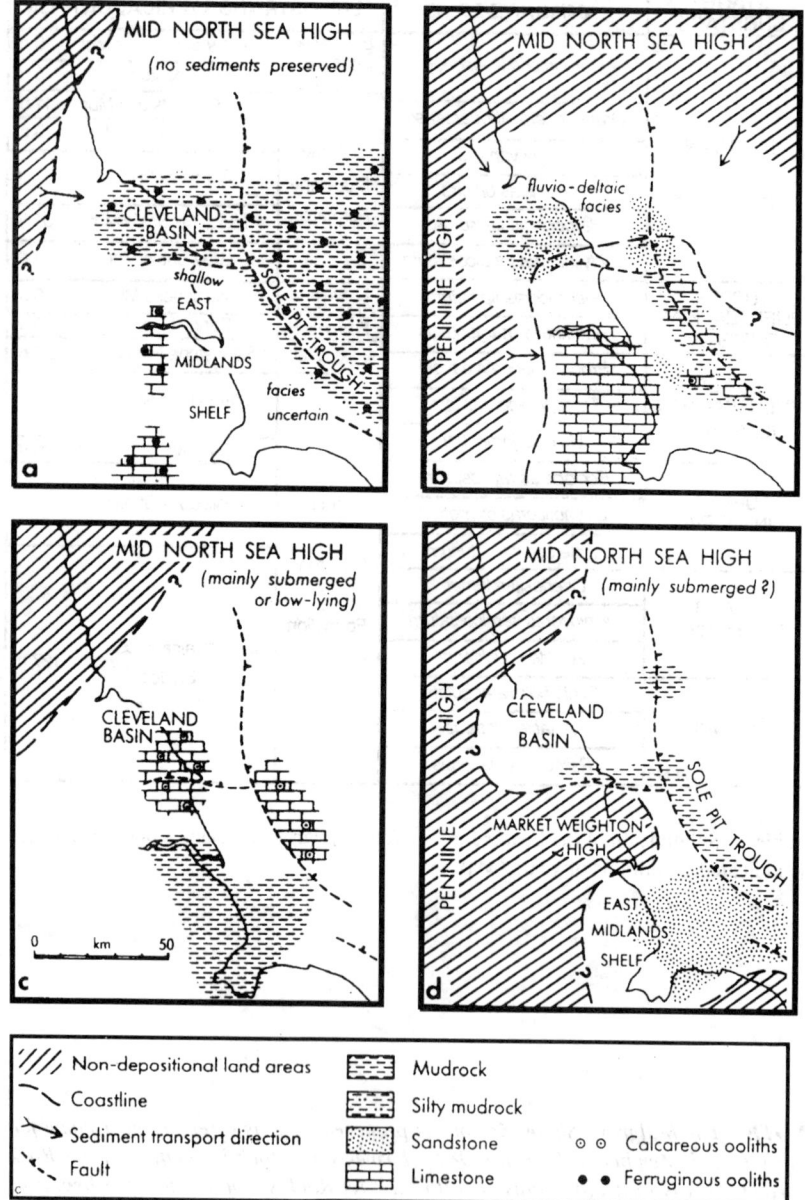

Figure 4. The palaeogeographical setting:
(a) Lower Jurassic   (b) Middle Jurassic   (c) Upper Jurassic   (d) Lower Cretaceous

argillaceous, deeper water succession composed of a series of minor coarsening-upward cycles on a similar scale to those of the Siliceous Shales. Each cycle is capped by an ironstone; the iron was presumably leached from a low-lying, subtropical, well-vegetated landmass.

The Toarcian saw important changes in the pattern of sedimentation in basins across Britain, reflecting a global rise in sea-level. This ushered in a new phase of mudrock sedimentation (Whitby Mudstone Formation). While the Grey Shales Member contains a normal marine fauna the overlying Mulgrave Shale Member (new name for the Jet Rock Member: see caption to Table I) shows evidence of oxygen depletion in the bottom waters, with a very restricted bottom fauna (Morris, 1979). It consists of dark, finely laminated shales sufficiently rich in hydrocarbons to smell of oil when freshly broken. The overlying Alum Shale Member marks a return to "normal" mudstone sedimentation, with some mud-borrowing bivalves.

Later Toarcian rocks are preserved only in small pre-Dogger (Middle Jurassic) synclines and are best exposed along the coast immediately to the southeast of the Peak Fault. Here, the shallower water sediments above the Alum Shale Member form a sequence of three fining-upwards cycles (the lowest forming the Peak Mudstone Member and the other two the Fox Cliff Member) overlain by two coarsening-upwards cycles (Grey Sandstone and Yellow Sandstone Members). Individual cycles vary between 9 and 12.6 m maximum thickness (Knox, 1984).

These phases of Late Toarcian shallowing may reflect successive early pulses in regional uplift and the creation of an extensive landmass over much of the central and northern North Sea. This culminated at the beginning of the Aalenian (Middle Jurassic). There then followed a regional sea-level fall and a radical change in palaeogeography (Figure 4b). Deltaic and fluviatile sediments emanating from the newly uplifted land area poured into the Cleveland Basin, in contrast to central and southern England which were covered by a warm, shallow shelf sea. At times seawater spilled into the basin from the south or east, so intercalating thin marine beds in the fluviodeltaic sequence.

The shallow marine Dogger Formation, which generally rests unconformably on early Toarcian sediments, is a complex unit, ferruginous throughout. Over much of the basin the formation is sandy, and in the present day coastal area is predominantly a tough, sideritic sandstone with a pebbly base. Towards the northwest, chamosite oolites and bioclastic limestones appear and a coastline probably lay to the north and west (Hemingway, 1974).

The depositional environment of the non-marine beds of the overlying Ravenscar Group (Table II) has caused much discussion. The initial estuarine hypothesis of Fox-Strangways (1892) has long been superseded and the debate now centres on deltaic versus alluvial plain environments. The group shows features characteristic of both. Coarsening-upwards cycles of bioturbated, laminated, intertidal siltstones with crevasse splay sheet sandstones, succeeded by rapidly filled distributory channels with slumps and water escape structures, suggest a deltaic origin. Meandering streams, superbly seen in the meander belt at Scalby Bay (Itinerary V), flood plains with desiccation cracks and footprint trails, lakes with beds of the bivalve *Unio*, marsh deposits, fossil soils with rootlet beds and the extensive development of sphaerosiderite all point to an alluvial origin.

The overall evidence suggests that following the marine intervals of the Dogger Formation, Eller Beck Formation, Lebberston Member and Scarborough Formation there was a rapid development of small prograding deltas, ultimately coalescing into a large alluvial plain; there is no evidence for a large single river system feeding into a large delta. The plain was always close to sea-level and susceptible to marine influence. The Scalby Formation shows the least marine influence, with a substantial thickness of the middle and upper Long Nab Member showing no sign whatsoever of marine microflora or of marine bioturbation.

While the marine incursion represented by the Lebberston Member probably came from the shelf to the south, the Eller Beck and Scarborough Formation transgressions were probably from the east (Knox, 1973; Parsons, 1977). The Scarborough Formation is firmly dated as mid-Bajocian, while the overlying Moor Grit and basal Long Nab members of the Scalby Formation have yielded very sparse dinoflagellate floras of uppermost Bajocian to Bathonian age. Dinoflagellates from the upper Long Nab Member at Newtondale indicate a Bathonian age (Riding & Wright, 1989). The evidence for a major break in the succession either beneath or above the Scalby Formation (Nami & Leeder, 1978) appears slim.

During the early Callovian the sea again transgressed into the Cleveland Basin from the east (Wright, 1977), the Cornbrash Limestone Formation resting on a bored erosion surface cut in Scalby Formation silts. Sea-levels continued to rise (following a global trend) to reach a local peak at about the Oxfordian/Kimmeridgian boundary. Since late Jurassic sediments are not preserved over most of the structural highs it is difficult to assess the distribution of land and sea, but facies patterns suggest that eventually most of the mid-North Sea High was flooded while the Pennine and Market Weighton Highs were at least partially submerged (Figure 4c).

STAGE	LITHOSTRATIGRAPHICAL DIVISION				
BATHONIAN 160 Ma	Ravenscar Group	Scalby Formation 65 m		Long Nab Member	
UPPER BAJOCIAN				Moor Grit Member	
— ? —					
		Scarborough Formation (m)			30 m
LOWER BAJOCIAN		Cloughton Formation 85 m		Gristhorpe Member	
				Lebberston Member (m)	12 m
				Sycarham Member	
— ? —					
AALENIAN		Eller Beck Formation			4 m
		Saltwick Formation			57 m
182 ma		Dogger Formation (m)			12 m

Table 2. Subdivision of the Middle Jurassic (Aalenian – Bathonian) sequence.

Callovian to Kimmeridgian sediments of the Cleveland Basin show significant facies differences to those on the East Midlands Shelf, though some of the changes took place away from the vicinity of the Market Weighton High and do not necessarily reflect instability of the high. Both the Kellaways Formation sands and much of the Oxford Clay Formation of the shelf are replaced north of the Vale of Pickering by a single sandy unit, the Osgodby Formation. This represents three phases of sedimentation separated by periods of erosion. For the first two phases the predominant source of sediment was to the northwest, though there is evidence of minor input from the east. Sediment incursion from these sources slowed down by the third phase, when sands of the preceding two phases were locally reworked. Wright (1978) has suggested that it was during the second phase (*jason* Zone) that the Market Weighton High was submerged to allow the Oxford Clay seas to spread clay northwards into the Vale of Pickering. By the beginning of the Oxfordian finer-grained sediments spread over the whole area to deposit a silty facies of the Oxford Clay Formation at least as far north as Scarborough.

Oxford Clay sedimentation did not last for long before the basin began to fill up and calcareous sandstones (grits) and limestones, including oolites and coral reefs, of the Corallian facies accumulated. These are equivalent to the highest Oxford Clay and much of the Ampthill Clay Formations of the East Midlands Shelf and indicate a second phase of inversion of basin and shelf. The Corallian facies in general indicate a warm, very shallow, well oxygenated sea in which carbonate banks and coral reefs developed. Fine-grained (micritic) limestones accumulated in the slightly deeper, backreef lagoons. The Lower Calcareous Grit, Coralline Oolite and Upper Calcareous Grit Formations constitute the Corallian (Table III) but the rapid lateral and vertical changes within them, especially in the Coralline Oolite Formaton, have resulted in a plethora of names for local members (summarised in Wright, 1972, 1983).

The Lower Calcareous Grit Formation is predominantly a *Rhaxella* (sponge) spiculite containing a small proportion of fine quartz sand, which was probably derived from the northern margin of the Market Weighton High (Wright, 1983). The siliceous spicules are readily dissolved and are the source of diagenetic cement which is often concentrated into chert nodules and bands. The settled, quiet conditions necessary for prolific growth of sponges seem to have suited ammonites very well, and the calcareous concretions of this formation yield in abundance some of the best preserved examples found in Britain.

Increasing wave energy, shallow shelf seas and minimal clastic input resulted in the deposition of substantial thicknesses of fossiliferous oolitic limestone. William Smith's term Coralline Oolite (Formation) is still used for these beds, corals occuring at many horizons. At three horizons near the base, middle and top of the formation the water cleared sufficiently for large patch reefs to develop. Some fine-grained sandstones occur too, the thickest being the Middle Calcareous Grit Member, which was derived from the north or northwest. Progressive uplift of the source areas brought an end to the reefs as red, lateritic clay was washed in. The clay is succeeded by siltstones, fine-grained sandstones and silicified *Rhaxella* spiculites of the Upper Calcareous Grit Formation.

In late Oxfordian times the Ampthill Clay facies spread northwards beyond Market Weighton to at least the Vale of Pickering. Clay sedimentation extended over the whole area by the beginning of the Kimmeridgian – a local reflection of a sea-level rise that led to a blanket of clay being deposited over much of the North Sea area from then on to the end of early Cretaceous times. The Kimmeridge Clay Formation is poorly exposed (Itinerary X) but sedimentation continued until early *pectinatus* Zone times. It then apparently ceased for about 9 million years (during which the sea probably remained over the area; Rawson & Riley, 1982) before commencing again with deposition of the Lower Cretaceous Speeton Clay Formation.

STAGE / SUBSTAGE			AMMONITE ZONE	LITHOSTRATIGRAPHICAL DIVISION
UPPER JURASSIC	VOLGIAN	PORTLANDIAN 140 Ma	Virgatopavlovia fittoni	
			Pavlovia rotunda	
			Pavlovia pallasioides	
			Pectinatites pectinatus	
		KIMMERIDGIAN	Pectinatites hudlestoni	
			Pectinatites wheatleyensis	
			Pectinatites scitulus	
	KIMMERIDGIAN		Pectinatites elegans	Kimmeridge Clay Formation c. 305 m
			Aulacostephanus autissiodorensis	
			Aulacostephanus eudoxus	
			Aulacostephanoides mutabilis	
			Rasenia cymodoce	
			Pictonia baylei	
		UPPER OXFORDIAN	Amoeboceras rosenkrantzi	Ampthill Clay Formation 46 m
			Amoeboceras regulare	
			Amoeboceras serratum	
			Amoeboceras glosense	Upper Calcareous Grit Formation 25 m
		MIDDLE OXFORDIAN	Cardioceras tenuiserratum	Coralline Oolite Formation 50 m
			Cardioceras densiplicatum	
		LOWER OXFORDIAN	Cardioceras cordatum	Lower Calcareous Grit Fm 42 m
			Quenstedtoceras mariae	Oxford Clay Formation 30 m
MIDDLE JURASSIC (pars)		CALLOVIAN	Quenstedtoceras lamberti	Hackness Rock Member 7 m
			Peltoceras athleta	Osgodby Formation — Langdale Member 18 m
			Erymnoceras coronatum	
			Kosmoceras jason	
			Sigaloceras calloviense	
		160 Ma	Macrocephalites macrocephalus	Redcliff Rock Member 25 m / Cayton Clay Formation 3 m
		(BATHONIAN)		Cornbrash Lst Fm 2 m
				(Ravenscar Group)

Table 3. Subdivision of the Middle to Upper Jurassic (Callovian – Volgian) sequence.

The Speeton Formation accumulated to the north of an emergent Market Weighton High (Figure 4d) which separated it from the more varied facies of the East Midlands Shelf. It crops out in a narrow strip along the Wolds escarpment to reach the coast south of Filey, where it is well exposed at the type section at Speeton (Itinerary X). There the clays are only about 100 m thick, but they provide a unique sequence through the British marine Lower Cretaceous, the lower (and better preserved) part of the section correlating with the non-marine Wealden facies of southern England. The faunal and floral successions provide a standard for comparison with the North Sea and North German successions. Thin volcanogenic mudstones at several levels reflect the activity of distant volcanoes, possibly in the Dutch sector of the southern North Sea.

The Speeton Formation passes up into a thin red limestone – the Hunstanton Formation or Red Chalk – deposited as rising sea-levels flooded the Market Weighton High again. As clastic input diminished a pure white limestone, the Chalk, began to accumulate, composed of calcareous algal plates (coccoliths) deposited as copepod faecal pellets in a clear warm sea that eventually extended over almost the whole of Britain and adjacent areas. The Chalk Group forms the arcuate hills of the Wolds, reaching the sea at Flamborough Head to form a magnificent sweep of cliffs about 17 km long. The lowest beds of the Chalk Group can be seen at Speeton (Itinerary X), but the overlying sequence in Buckton and Bempton cliffs is inaccessible. However, the famous contortion zone at Scale Nab (Old Dore and Little Dore) is readily seen from pleasure boats and together with the structure in Selwicks Bay (Itinerary XI) forms part of the Howardian-Flamborough Fault Belt. In the Flamborough region the beds are accessible again (Itineraries XI to XIII).

Since the previous edition of this guide was written there has been a major revision of the stratigraphy of the Chalk of the northern province (Yorkshire to North Norfolk). The former division into Lower, Middle and Upper Chalk was unsatisfactory and the long established fossil zones were very vaguely defined. Recognition of the lateral continuity of flint and marl bands led Wood and Smith (1978) to propose a lithostratigraphical division in which the Chalk Group embraces four formations (Table IV), with numerous named marker horizons in the lowest three units. The Flamborough Formation has now been subdivided too (Whitham, in press). Thus there is a firm framework against which fossil occurrences can be calibrated (Whitham, 1991 and in press). Much of the sequence is demonstrated in Itineraries X-XIV.

No Tertiary sediments are known from the area, although sands and clays preserved in solution hollows in the Chalk of the Wolds were considered by Versey (1939) to be of pre-glacial (i.e. Tertiary) age. The burial history of the Cleveland Basin also indicates that some Lower Tertiary sediments may have been deposited, then removed during basin inversion later in the Tertiary.

THE YORKSHIRE COAST

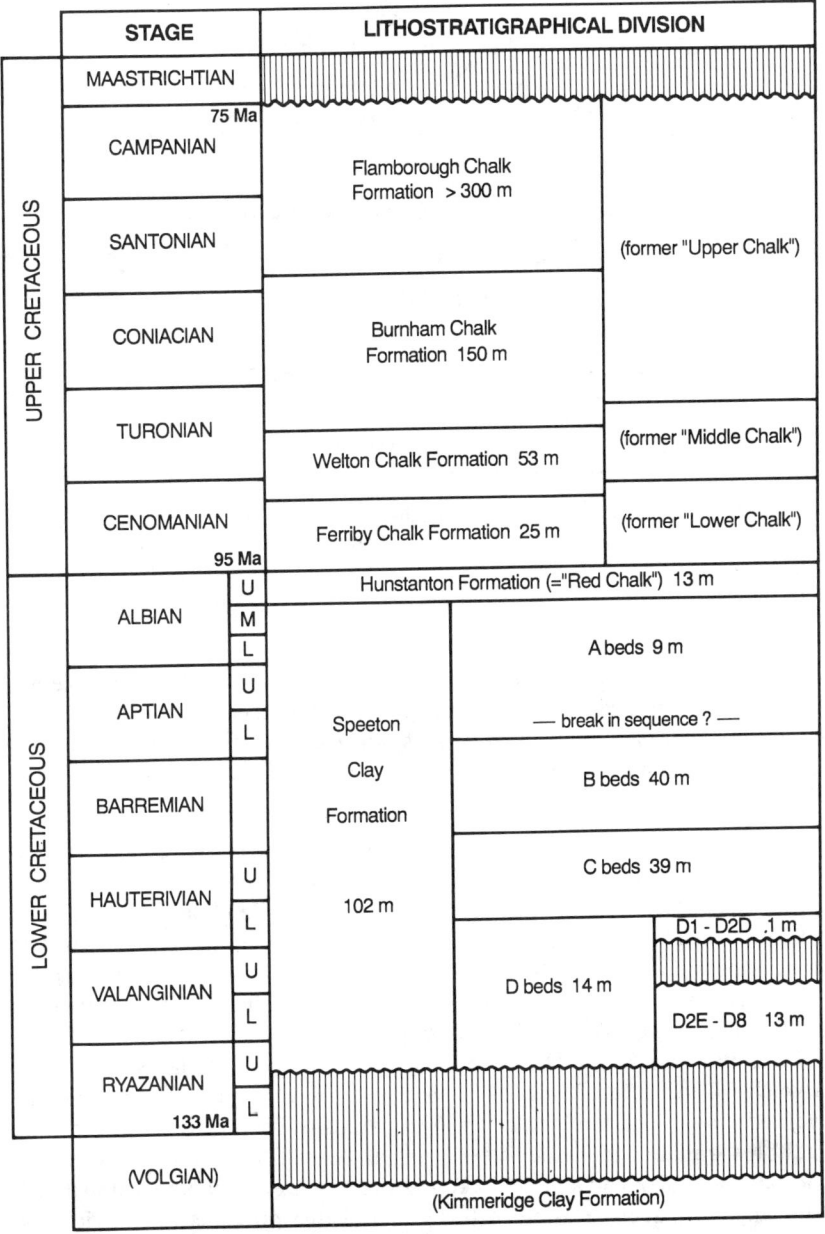

Table 4. Subdivision of the Cretaceous (Ryazanian – Campanian) sequence.

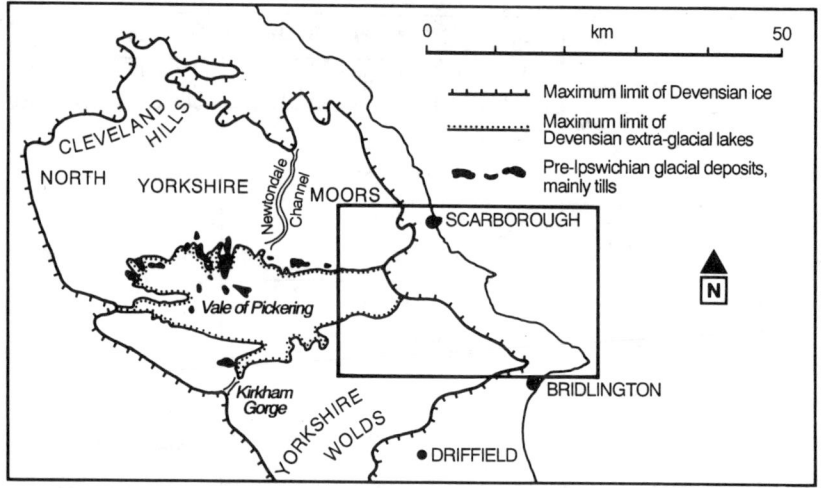

Figure 5. Geography of northeast Yorkshire during the last (Devensian) glaciation (modified from Kent, 1980b, fig. 27). The rectangle indicates the area shown in detail in Figure 37.

Tertiary igneous activity is represented by the Cleveland Dyke, a tholeiitic basalt intrusion peripheral to a volcanic province centred on the western Scottish island of Mull. The dyke does not reach the coast but is visible on the North Yorkshire Moors (Itinerary VI). It has been radiometrically dated to 58.4 ± 1.1 Ma (i.e. Eocene) (Kent, 1980b, p. 107).

During the Pleistocene ice age the area was glaciated at least twice, though only a few relics remain of deposits pre-dating the last (Devensian) glaciation. During the Devensian much of the higher land probably remained ice-free, but glaciers passed down the Vale of York and also abutted against the northern flanks of the North Yorkshire Moors and the coast (Figure 5). Ice also infilled the pre-glacial bays. A system of meltwater lakes filled the valleys and flowed via overflows channels into the Vale of Pickering (Figure 37 and Itinerary XIV). The main deposits left behind are the boulder clays (which fill the core of each bay), sand and gravel moraines and lacustrine varved clays. The boulder clays are composite sheets containing erratic pebbles and boulders derived from at least three distant sources – Scandinavia, Scotland and the Lake District – as well as more local material. It is the subsequent erosion of this material that provides such a rich and fascinating suite of pebbles on our beaches – including the carnelians and agates so popular with collectors.

ITINERARY I

Staithes to Port Mulgrave

P. F. Rawson

O.S. 1:25,000 Outdoor Leisure Map Sheet 27 or
1:50,000 Sheet 94 Whitby
G.S. 1:63,360 Sheet 34 Guisborough

The 3 km stretch of coastline between Staithes and Port Mulgrave provides a magnificent series of exposures of the Lower Jurassic Staithes Sandstone, Cleveland Ironstone and lower Whitby Mudstone Formations. The rugged splendour of this coastline has been much modified by man; the former working of ironstone, alum shale and jet in the area has left extensive evidence, some of which is mentioned below. Now, the skyline to the west of Staithes is dominated by the Boulby potash mine.

The route is mainly over a rocky wave-cut platform and **a falling tide is essential** as the rising sea reaches the foot of the cliff in places by about mid-tide.

Locality 1. Staithes to Penny Nab (Staithes Sandstone Formation).

Staithes is a picturesque fishing village which retains much of its original character – though many of the cottages are now second homes. From the village car park (Figure 6) walk down the main road, noting to the left the 30 m deep gorge cut through the Staithes Sandstone Formation by the post-glacial Staithes Beck. Continue through the older part of the village to the harbour wall (NZ 784188), where the Staithes Sandstone Formation (28.6 m thick) can be seen in the cliffs on both sides of the harbour. Howarth (1955) showed that the lower part of the formation belongs to the *davoei* Zone (12.6 m thick according to Howard, 1985) and the upper part to the *margaritatus* Zone (16 m thick). It consists of shallow marine sandstones and siltstones with clearly displayed sedimentary structures. The thicker bedded units are often cross-bedded; they include excellent examples of hummocky cross-stratification, which is believed to have formed through the reworking of shallow sands by oscillating storm waves. Thinner bedded units consist of sheets of fine sandstone fining-upwards to mudstone, the base of individual sheets being erosive. They show delicately preserved parallel lamination, low-angle cross-lamination and wave ripple lamination, but in many cases this is at least partially destroyed by bioturbation which often becomes intense in the more argillaceous upper part of each sheet.

The lower part of the formation is well exposed at Cowbar Nab, to the west of the harbour, while the higher beds are most accessible to the east.

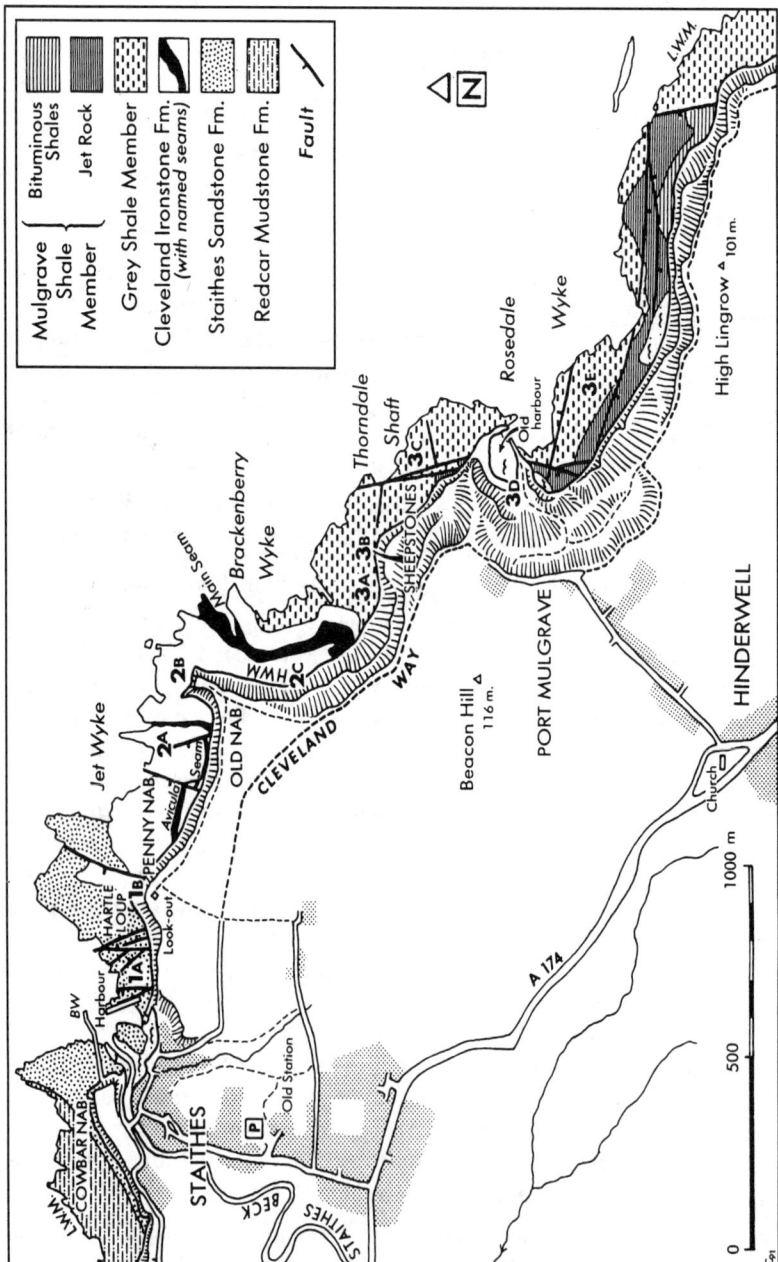

Figure 6. Map of the Staithes – Rosedale Wyke shore (based mainly on Howarth, 1955, 1962, 1973).

1A. Eastern side of Staithes Harbour. The higher part of the Staithes Formation forms the lower part of the cliff and adjacent scars. Higher in the cliff the individual ironstone bands of the Cleveland Formation stand out clearly. In the vertical faces at the cliff foot all the sedimentary features noted above can be seen. On adjacent scars ripple-marked surfaces are visible and sideritic concretions preserve large numbers of the bivalves *Protocardia truncatum*, *Oxytoma cygnipes* and *Gryphaea depressa*, and the scaphopod *Dentalium giganteum*.

Several minor faults occur here, especially in a small recess in the cliff known as Hartle Loop, about 150 m east of the harbour.

1B. Penny Nab. Eastwards towards Penny Nab the Staithes Formation becomes increasingly argillaceous upwards until it grades into the cyclic sediments of the overlying Cleveland Ironstone Formation. The base of the latter formation (and of the Penny Nab Member) is taken at the base of the first cycle (Howard, 1985), i.e. at the base of Howarth's (1955) bed 24. This is a row of scattered siderite mudstone nodules, sometimes packed with small ammonites, occurring round the foot of Penny Nab at the base of a sloping ledge. A few paces to the northeast of this ledge a series of parallel grooves in the shale at regular 1.22 m (4 feet) intervals marks the line of an old tramway for transporting ironstone to a shallow dock northwest of the Nab (Owen, 1985, figure 2).

Locality 2. Jet Wyke to Brackenberry Wyke (Cleveland Ironstone Formation).

In the type area of the Cleveland Hills the Cleveland Formation contains thick ironstone seams which were formerly mined extensively. The ironstones thin and the intervening shales thicken towards the coast, where the formation is excellently exposed in Jet Wyke and round Old Nab into Brackenberry Wyke. Here, it consists of 25.3 m of shales and thin siltstones with sideritic and chamositic ironstone seams, some of which are oolitic in texture. Most seams cap coarsening-upwards cycles up to 7 m thick (Figure 8), and individual cycles are laterally continuous over much of the basin (Rawson, Greensmith & Shalaby, 1982; Howard, 1985). The upper part of some cycles is striped with thin fining-upwards sheets, sometimes with basal gutter marks, probably deposited under storm conditions ("tempestites"). The formation is divided into the predominantly shaly Penny Nab Member (18 m) and the more ferruginous Kettleness Member (7.3 m) (Howard, 1985).

2A. Jet Wyke. The whole of the Cleveland Formation is accessible here, though the highest beds are better examined around Old Nab. The succession dips gently eastwards and several faults repeat parts of it, in one case bringing ironstone against ironstone in the middle of the Wyke to create a very extensive ironstone pavement on the shore. A detailed lithological sequence was given by Howarth (1955) and this has been combined with more recent sedimentological work to show the whole sequence as a log (Figure 7). The lowest ironstone, the Avicula Seam, forms a flat ledge starting about 150 m east of Penny Nab. Its upper surface shows many specimens of *Oxytoma* (formerly *Avicula*) *cygnipes*, while the base is conglomeratic.

Further east, the "upper striped bed" of Greensmith, Rawson and Shalaby (1980) is often cleanly exposed at the cliff foot immediately beneath the thin (10 cm) Raisdale Seam ironstone. This bed (about 2 m thick and forming the upper part of bed 34) consists of a series of delicately preserved layers of pale coloured, laminated siltstone fining-up to darker mudrock (Figure 8). Each layer has an erosive base and gutters are developed at the base of some, cutting down into, and even undercutting, up to six underlying layers. The bed can be traced from the cliff foot onto the scars towards Old Nab, where the lighter, hard bases of the anastomosing gutters are seen in plan to be up to 0.5 m wide and 5 m long with an orientation almost due east-west. Similar gutters in the same bed at Hawsker (20 km to the southeast) are much less deep and have a finer-grained infill as if they are more distal from the shoreline, indicating currents from the west. The sequence at both localities suggests deposition under storm surge conditions.

Along the eastern side of the wyke the Pecten Seam is seen in vertical section to be represented by five thin layers of ironstone separated by thin shale partings. The base of the seam is taken as the base of the Kettleness Member.

2B. Old Nab. The regular blocks at the tip of this prominent headland are the result of former ironstone mining; pillars of Main Seam ironstone left as roof supports have since been unroofed by marine erosion. On the east side of the Nab shale backfill can be seen filling an adit running into the cliff. In the immediate vicinity of the Nab and into Brackenberry Wyke the Main Seam bedding surfaces show extensive networks of *Rhizocorallium* (crustacean) burrows, most of them showing scratch marks made by the crustaceans' claws.

2C. West side of Brackenberry Wyke. Here, the higher part of the Cleveland Formation is gently arched so that the lowest beds are seen about halfway along the cliff. Mine adits are again visible and ironstone was also quarried from the shore. There are good exposures of bedding surfaces of the Pecten and Main Seams, the former crowded with *Pseudopecten equivalvis*

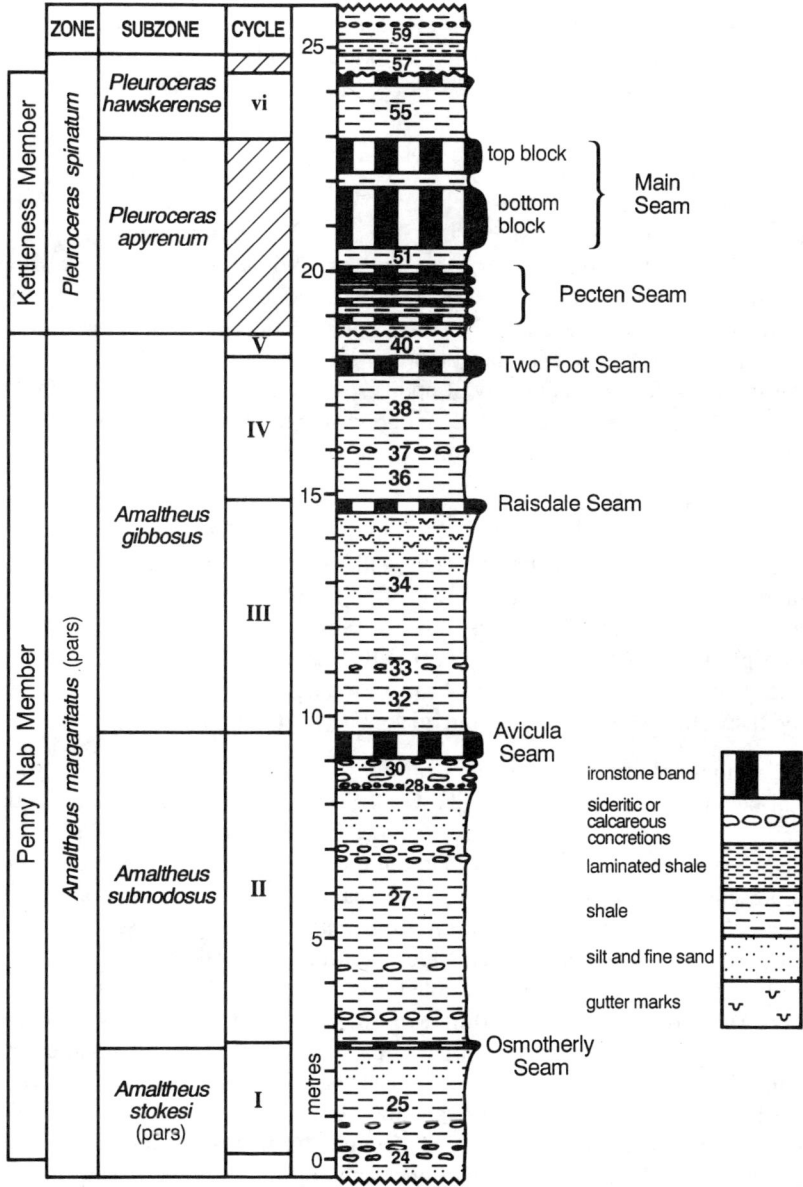

Figure 7. The Cleveland Ironstone Formation sequence at Staithes (modified from Howarth, 1955, and Howard, 1985).

and the latter riddled with *Rhizocorallium*. Towards the head of the Wyke the ironstone nodules of bed 56 form the last ironstone platform. They contain body chambers of *Pleuroceras hawskerense* plus numerous bivalves (*Pleuromya costata* and *Gresslya*) still preserved in burrowing position.

Locality 3. Brackenberry Wyke to Port Mulgrave (Whitby Mudstone Formation).

3A. Southwest side of Brackenberry Wyke. Just above bed 57 of the Cleveland Formation a sharp facies change marks the basal Toarcian transgression. The Grey Shales Member of the Whitby Mudstone Formation consists of grey micaceous mudstones with concretionary horizons. The lower part contains six bands of red-weathering sideritic concretions which cross the rocky platform at the south end of the Wyke. Here, the shingle on the shore above consists largely of nodules derived from various levels in the Whitby Mudstone Formation and is a good source of ammonites. It is better to collect here than to remove *in situ* specimens.

3B. The Sheep Stones. On the southern corner of Brackenberry Wyke the shore is strewn with blocks of sandstone that have fallen from the Hayburn Formation at the top of the cliff. On walking across this area note that the more easterly blocks in particular are often firmly embedded on shale stacks up to half a metre high. These are the Sheep Stones, which have been interpreted to represent an ancient fall that took place when the rocky intertidal platform and mean sea-level were slightly higher, probably during the Ipswichian (= penultimate) interglacial period (Agar, 1960).

3C. Thorndale Shaft. The shales on the shore to the southeast of the rock fall belong to the upper part of the Grey Shales Member. The characteristic ammonite *Dactylioceras tenuicostatum* occurs mainly in nodules which have been extensively collected, so that some of the principal nodule beds are represented by lines of hollows in the scars. The highest beds contain *D. semicelatum*, joined in the top 1.8 m by *Tiltoniceras antiquum* (see section in Howarth, 1973). Towards the cliff foot the Mulgrave Shale Member outcrops and forms the whole of the accessible lower part of the cliff between here and Port Mulgrave harbour. The Mulgrave Shale Member is a finely-laminated, richly pyritic, dark-coloured, bituminous shale with numerous concretions either in beds or randomly scattered. These are usually pyrite-skinned, very hard and splintery and therefore **very dangerous to hammer.** There are four main marker beds, the Cannon Ball Doggers (which form the base of the member), Whale Stones, Curling Stones and Top Jet Dogger (Howarth, 1962, gives a detailed section). A mine adit about half-way along the cliff provides a good reference point; the thin (25 cm) stone band forming the roof of the adit is the Top Jet Dogger.

Figure 8. The Cleveland Ironstone Formation in Jet Wyke. *(Upper)*. The ironstone pavement in the foreground is the Avicula Seam. The shales above coarsen upwards into a more silty sequence, the "Upper Striped Bed", which is capped by the Raisdale Seam (R) across the middle of the photograph. The next hard band is the Main Seam, followed by the five thin ironstone ribs forming the Pecten Seam (P), visible at the top of the photograph. *(Lower)*. Close-up of the upper part of the "Upper Striped Bed", showing fining-upwards sheets with erosive bases. Well developed gutters are particularly prominent at the level indicated by the lens cap (3 cm diameter). The base of the overlying Raisdale Seam ironstone is conglomeratic.

The fauna of the Mulgrave Shale Member consists mainly of free-swimming animals (ammonites, fish, reptiles, etc.) and bottom-living organisms are generally absent, though the specialised bivalve *Pseudomytiloides dubius* is common. The main water mass must therefore have been well oxygenated while the sea floor was oxygen-deficient. There was a widespread "anoxic event" at this time and such shales were deposited over much of Europe (Posidonienschiefer, Schistes Carton, etc.). They form a source rock for oil in North Germany, the Netherlands and the Paris Basin, and when freshly broken the Yorkshire sediments smell strongly of oil. The lower part of the member ("Jet Rock" *sensu stricto*) is more finely and persistently laminated than the higher part ("Bituminous Shales") and represents peak anoxic conditions (Pye & Krinsley, 1986).

Jet occurs as long, compressed seams, lustrous black in cross-section. It is most easily found after winter storms when cliff falls have occurred. Later in the year the former position of seams is marked by small excavations in the cliff!

3D. Port Mulgrave. Port Mulgrave harbour was built in 1856-7 to ship ironstone worked from the shore and adjacent mines to the iron works of Tyneside. The bricked-up entrance to main mine roadway (Seaton Drift) is visible in the cliff. Plans of the mining area and detailed information of the history of the mining are given by Owen (1985).

3E. Rosedale Wyke. The Grey Shale and Mulgrave Shale Members form the rocky platform here (Figure 6), and show the same marker beds seen in the previous exposure. The Jet Rock was formerly quarried from the shore and the main marker beds are readily traced. Note in particular the Curling Stones, large rounded concretions that the jet miners left as stacks on the shore (Hemingway, 1968, p. 10). Above are the Millstones, lenticular doggers of limestone up to 3 m in diameter set in the upper surface of the Top Jet Dogger; these occur close to the cliff foot in the middle of the Wyke.

Return from here to the adjacent harbour where a path up the cliff leads to a footpath back to Staithes. Alternatively take the narrow road (where there is parking for a car or minibus) leading to the main road (A174) at Hinderwell. This is the shorter route for a coach pick-up, as a coach can park on the grass just off the main road, near the church, at NZ 791171.

ITINERARY II

Whitby to Saltwick Bay

J. E. Hemingway, revised by J. K. Wright

O.S. 1:25,000 Outdoor Leisure Map Sheet 27 or
1:50,000 Sheet 94 Whitby
G.S. 1:63,360 Sheet 35 with 44 Whitby and Scalby

This short excursion examines the Whitby Mudstone Formation and part of the Ravenscar Group. It follows the rock platform at the base of the cliffs from Whitby to Saltwick Bay, ascending the grassy cliffs in the bay and returning to Whitby via the cliff top path (Figure 9). The going is easy. **The tide should be falling and at least half way out before commencing.** Difficulties could be experienced on a rising tide as the sea eventually reaches the base of the cliff. Parking is available on the harbour-side car park on the opposite side of the harbour, otherwise drive up to the Abbey car park and descend to the shore via the Abbey steps.

Locality 1. Whitby Pier (Alum Shale Member and the Whitby Fault).

Proceed north along Church Street, and at the Board Inn turn right uphill, left at the foot of the Abbey steps, and along the footbridge leading to the East Pier. From the foot of the East Pier Lighthouse, contrast the succession of the East Cliff (52 m thick) with that of the West Cliff (23 m thick). To the east, the rock platform and lower cliff expose 12 m of Alum Shales. These are succeeded by a thin representative of the Dogger Formation (0.75 m) which is in turn followed by the fine sandstones, siltstones and carbonaceous clays of the Saltwick Formation (31.4 m). There follows the marine Eller Beck Formation (c. 6 m), which is capped by boulder clay (Figure 10). The Alum Shale Member may be traced, at low tide, within the harbour at Collier Hope and also to the east of the West Pier.

By contrast, the West Cliff and Khyber Pass reveal a solely non-marine sequence of channel sandstones stacked one above the other. Because of this striking difference, a NNE/SSW fault with a westerly downthrow of 61 m was postulated along the deep channel of the harbour. The West Cliff succession was thus assigned to the Cloughton Formation in the middle of the Ravenscar Group (Fox-Strangways & Barrow, 1915). However, subsequent detailed mapping, particularly of the Dogger Formation, as well as borehole evidence within the harbour, shows that the fault (which is not exposed) downthrows no more than 12 m to the west. The West Cliff succession is thus a part of the Saltwick Formation, a conclusion which has been confirmed by

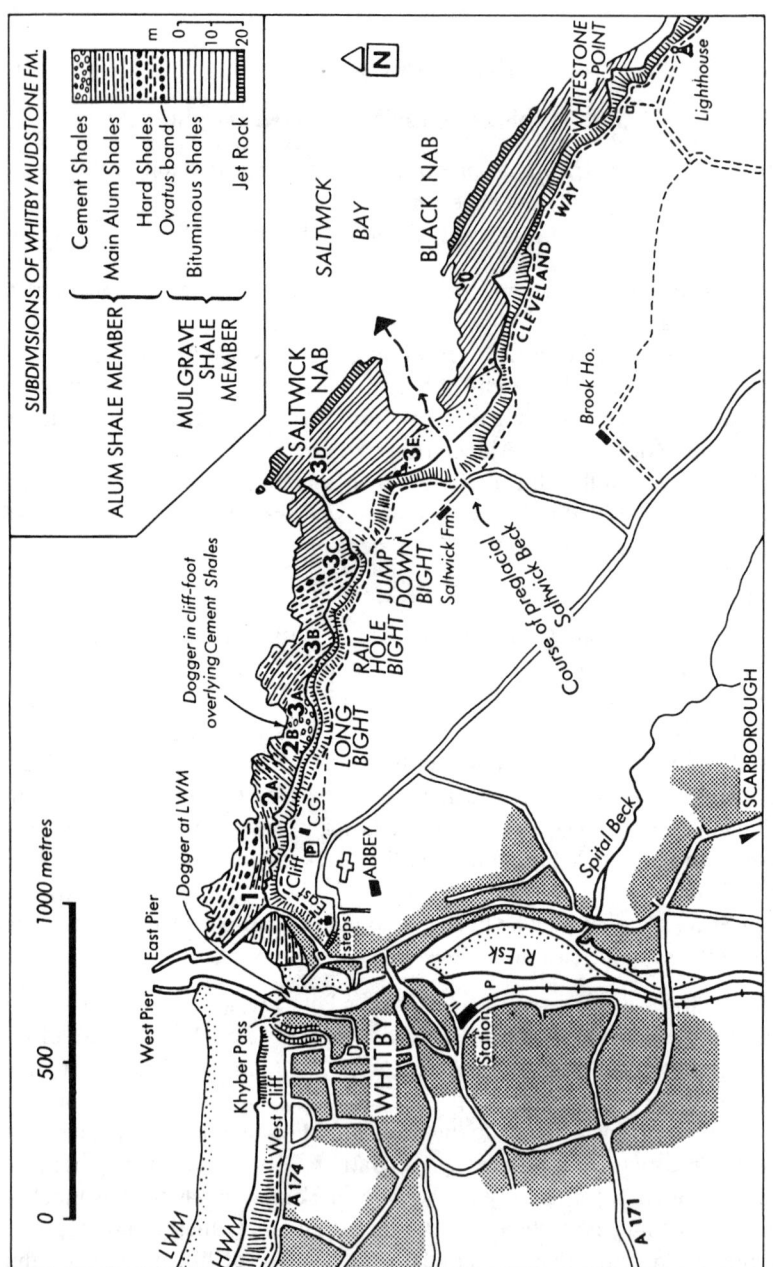

Figure 9. Map of the Scaur, Whitby. *Note*: roads are shown only where relevant to the itinerary.

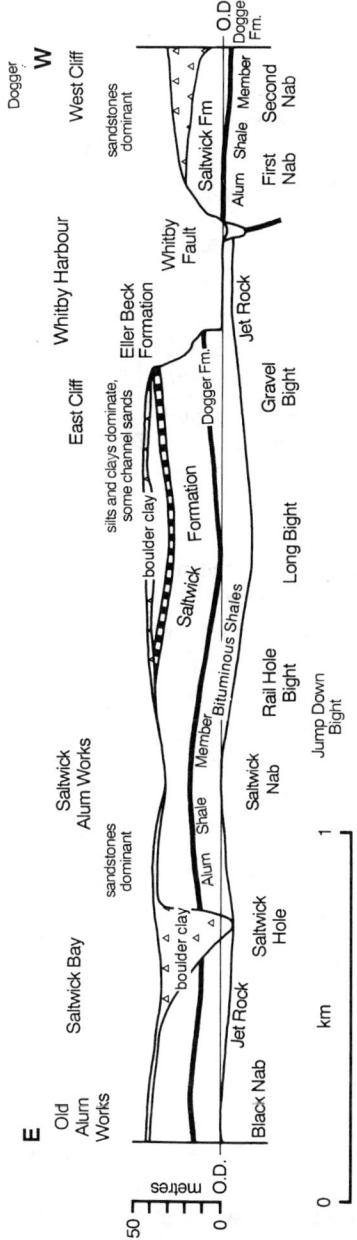

Figure 10. Cliff section at Whitby, as seen from the shore.

spore analysis. Alexander (1986) has proposed that the Whitby Fault was active in the Middle Jurassic, and that on the downthrown side to the west was a persistent, low-lying area repeatedly occupied by river channels. This would explain the succession of infilled channels visible to the west, whereas the Saltwick Formation to the east is largely level-bedded.

Return along the East Pier and descend to the shore by the steps on the harbour side just before the bridge. Pass under the concrete bridge via the steps in the concrete breakwater, keeping close to the pier, and descend to the rock platform or 'Scaur'. The shore from Whitby to Saltwick Bay and beyond is made up entirely of shales of the Whitby Mudstone Formation (Powell, 1984), this being the type-section. Because of the gentle folding, beds from the Jet Rock to the Cement Shales may be examined.

The Alum Shale Member near the East Pier is rich in the thick-shelled bivalve *Dacryomya ovum*, both crushed and uncrushed, which occasionally is sufficiently abundant to form thin limestones. Some *Dacryomya* are in life position, as also are *Pleuromya* sp. The ammonites *Dactylioceras* and *Hildoceras* are common. Many, however, have only the outer whorl (body chamber) well preserved, the inner whorls being crushed within a calcareous mudstone concretion.

Notice the strongly developed north-south jointing, both vertical and inclined, which controls the cliff recession. All stages of cliff erosion from the widening of joints to the formation of vertically walled caves, capped by the Dogger Formation, may be seen. To the east, scattered fragments of wood up to 2 m long occur in the shales. They are thickly coated with pyritic mudstone, the product of anaerobic decay, and may show either well-preserved cellular tissue infilled with calcite or barytes, or crushed and carbonised woody tissue preserved as poor-grade jet.

Locality 2. Whitby cliffs (Dogger and Saltwick Formations).

2A. East of Whitby. Rounding the first headland east of Whitby, a large rock fall (c. 1912) is seen 200 m ahead. Approaching the rock fall, note that the Dogger Formation in the cliff has been thinned almost to nothing by erosion at the base of a broad channel in the overlying Saltwick Formation. The western side of the channel is filled with gently dipping sandstone/siltstone alternations. Eastwards, a large incursion of sand into the channel system is marked by thick, easterly dipping cross-sets. These were laid down on the inside of a bend, as a point-bar, as the channel migrated eastwards. On the eastern side of the rock fall is an excellent section near the cliff foot showing coarse-grained, cross-bedded channel sandstone containing pebbles and carbonised tree trunks, resting on the Dogger Formation. Fifty metres to the east the channel is absent. Almost certainly it became choked with tree trunks and sediment, and was abandoned.

2B. Long Bight cliffs. The small bay immediately east of the rock fall is known as Long Bight. In the centre of the bay the Dogger Formation reaches beach level, and is overlain by thin beds dipping eastwards. These are collectively about 3.5 m thick and are distinctive in lithology, being fine-grained, pale grey sandstones with concretions of sideritic mudstone. They exhibit much contemporaneous microfaulting. They yield a flora including *Coniopteris, Williamsonia, Baieria* and *Czechanowskia*, and are known as the Whitby Plant Bed. This is evidently an offbank deposit laid down under quiet conditions. The relationships of this deposit to the channel, to the underlying Dogger Formation and to contemporaneous small-scale faulting are obscured at present (1988) by the accumulation of debris at the cliff foot.

The Dogger Formation here is a tough, sideritic sandstone, 0.4 m thick, forming a distinctive shelf rising gently up the cliff. Twenty metres east of the plant bed abundant U-shaped tubes allied to *Arenicolites* or *Diplocraterion* descend from the base of the pebbly Dogger Formation into the Alum Shale Member. Four metres above the Dogger, a bed, dominated by thick horizontal stems of carbonised wood, has been worked for coal. Blocks of a fine grained sandstone fallen from a horizon 5.7 m above the Dogger Formation and which carry paired valves of *Unio kendalli*, sometimes gaping, are equivalent in age to the freshwater shell-bed originally described at Saltwick (Jackson, 1911).

Locality 3. The shore between Long Bight and Saltwick Bay (Whitby Mudstone Formation).

3A. Long Bight shore. On the shore a gentle syncline (Figure 10) has brought down the upper part of the Alum Shale Member, known as the Cement Shales, to shore level. These shales are characterised by abundant concretions of calcareous mudstone, at one time burnt at Sandsend for hydraulic cement. They yield *Catacoeloceras* and less commonly *Hildoceras bifrons* and *H. hildense*. There are abundant belemnites, mainly *Salpingoteuthis*.

3B. Rail Hole Bight. Rounding the next headland eastwards, a small bay known as Rail Hole Bight is reached. The Dogger Formation climbs the cliff with a largely channel-free succession above it. Progress eastwards is now down the succession. At the foot of the headland, on the opposite side of Rail Hole Bight, a sideritic ironstone band within the Alum Shale Member runs into the base of the cliff. Twenty metres seawards of the point, a second 0.2 m sideritic mudstone separates the Main Alum Shales from the Hard Shales below.

3C. Jump Down Bight. Entering Jump Down Bight, the bay just before Saltwick Nab, proceed across the Hard Shales to the southernmost corner of the bay. Fifty metres out towards Saltwick Nab in the base of the cliff is a double row of pyrite-skinned concretions associated with lenses of impure limestone, showing cone-in-cone structure, and with sideritic mudstone. This is the Ovatum Band, yielding occasional *Ovaticeras pseudovatum*. It is the highest division of the Mulgrave Shale Member.

3D. Saltwick Nab. Continue down the succession across the Bituminous Shales (Mulgrave Shale Member), round the foot of Saltwick Nab, and observe numerous squashed harpoceratids in the shale. Near the end of the nab the Bituminous Shales yield abundant squashed *Pseudomytiloides dubius*; on the surface of concretions they occur uncrushed. The low water reef is made of the tough, impure limestone (the Top Jet Dogger) at the top of the Jet Rock. This limestone is characterised by discoidal concretions known as the Millstones, which may be as much as 4.2 m in diameter, and which are set in the top surface of the limestone. Below it, the Jet Rock, a finely-laminated black or black-brown shale, rich in pyrite-skinned concretions, has been worked extensively for jet here, even though it is accessible only at low water spring tides, and is not normally seen. Freshly broken Jet Rock smells strongly of bitumen as do the overlying Bituminous Shales, and oil may be found in the camerae of those ammonites which are preserved in concretions. Fragments of good quality 'hard' jet, a hard coaly substance formed by the alteration of plants such as conifers allied to araucarians, are only rarely found.

3E. Saltwick Bay. The Ovatum Band occurs near or at the foot of the cliff within the north end of Saltwick Bay. In addition to *Ovaticeras*, it has yielded large masses of belemnite-limestone about 7 cm thick. Above, the bay is marked by a large overgrown quarry in Alum Shale south of Saltwick Nab. It was excavated for the making of alum, as was a smaller quarry south of Black Nab. Heaps of red, burnt shale, overgrown soaking-pits and the ruins of old quays, as well as the size of the quarries here and elsewhere in this region, are an indication of the importance of this industry, which flourished during the sixteenth to nineteenth centuries, and whose history is described by Fox-Strangways (1892, pp 452-5).

Ascend the cliff by the track above the sandy beach. Here, a pre-glacial channel with a near-vertical west bank and a gentler eastern slope is entirely filled with boulder clay. It persists seawards as a deep channel (Saltwick Hole) which is particularly noticeable at low tide. In the upper part of the cliff the Saltwick Formation is dominated by massive sandstones in striking contrast to the succession in the East Cliff at Whitby.

From the cliff-top contrast the form of Saltwick Nab with that of Black Nab. The former has been rapidly eroded since 1940, after a period of relative stability before the war, and will be reduced to a stack within a short time. Black Nab is now only a stump of shale, the remaining pedestal of three stacks which were eroded in the storm surge of January 1953.

Return to Whitby via the cliff-top path.

ITINERARY III

Robin Hood's Bay

J. K. Wright

O.S. 1:25,000 Outdoor Leisure Map Sheet 27 or
1:50,000 Sheet 94 Whitby
G.S. 1:63,360 Sheet 35 with 44 Whitby and Scalby

It is not practicable to produce an itemised itinerary to these classic exposures of the Redcar Mudstone Formation (Lower Lias) in the manner of the other itineraries in this volume. Robin Hood's Bay is large (6 km from North Cheek to Peak Steel) and individual beds may be traced for up to 2.5 km round the bay, so that it would be difficult for the visitor to find specific localities mentioned here. Instead, Figure 11 gives a plan of the geology of the bay, indicating car parks and access roads and paths. The Redcar Mudstone Formation should be examined on the rock platform, **not at the cliff foot,** as the crumbling shale cliffs can fall at any time and severe injuries have been caused by falling blocks of shale in recent years. There is the very real prospect of being cut off by an incoming tide, which eventually reaches the base of the cliff right round the bay. Holiday makers have to be rescued by boat every year. Therefore, any circuit of the bay should be done only on a falling tide. The going is easy.

The bay is accessible from four directions (Figure 11). From the south side, one can enter or leave by the route from Ravenscar to the beach described in Itinerary IV. There is a regular bus service from Scarborough to Ravenscar. Alternatively, one can proceed over Stoupe Brow to Stoupe Beck, where a broad track leads to the beach. This road is narrow and the car park small, so that only cars or minibuses are suitable. The same is true for the larger car park at Boggle Hole, accessible from the Scarborough-Whitby road to the southwest. Robin Hood's Bay Town, with its car parks and shops, is accessible from the northwest. Buses from Scarborough and Whitby regularly visit the town. One possibility which a party with limited time in the area might wish to consider would be to drive to Stoupe Beck, complete Itinerary III working southwards, and then Itinerary IV, arranging for the driver(s) to bring the vehicle(s) round to Ravenscar.

Robin Hood's Bay is famed for displaying a dome-like structure with its eastern and central parts lying below sea-level (Figure 14). What remains has been carved by erosion into a huge half-amphitheatre, revealing in the cliffs and rock platform much of the Sinemurian and the whole of the Lower Pliensbachian sequences. The only Lower Liassic stage not exposed is the

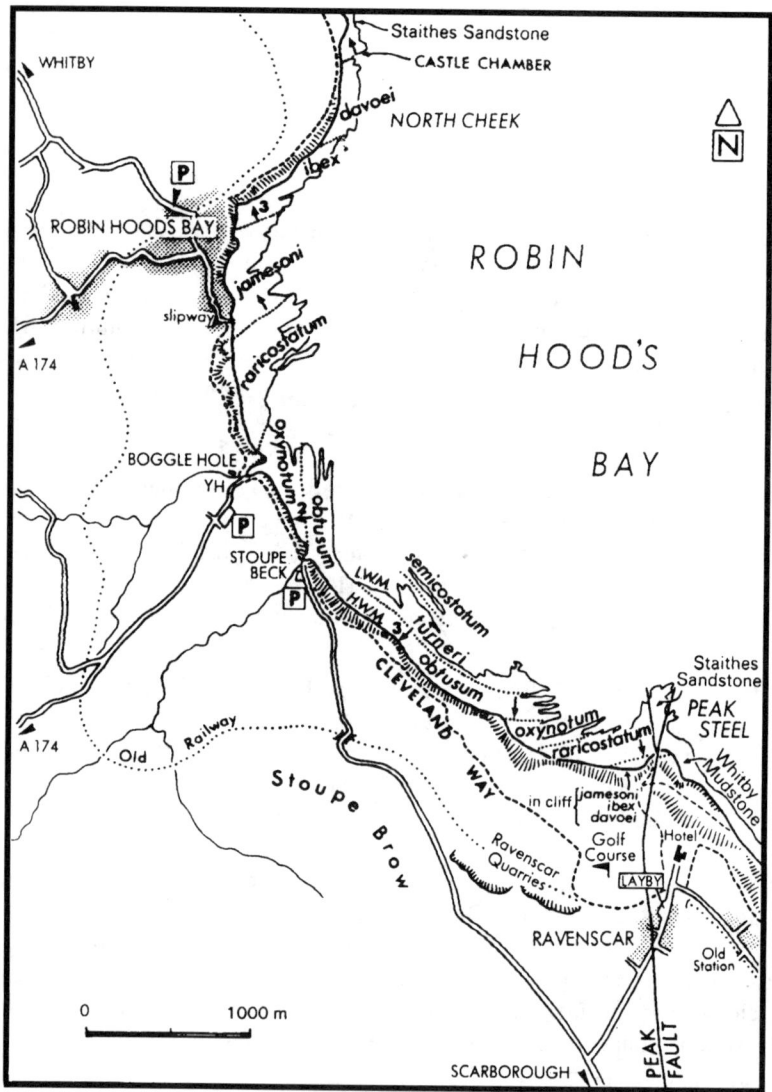

Figure 11. Geological sketch map of the Robin Hood's Bay area.

Hettangian, some 90 m of which was met in nearby boreholes. Using the informal lithological subdivisions introduced by Buckman (1915), and followed by Bairstow (1969), the succession (see also Table I) is:

Ironstone Shales	56.5 m
Pyritous Shales	20 m
Siliceous Shales	40 m
Calcareous Shales seen to	20 m

The Calcareous Shales are exposed in the low tide reefs in the centre of the bay. They consist of silty, calcareous mudstones with frequent calcareous concretions. The characteristic ammonites *Arnioceras* and *Caenisites*, with chambers filled with calcareous mudstone, can be found by careful search of the seaweed-covered reefs.

The Siliceous Shales swing round in a huge arc from Boggle Hole nearly to Peak Steel. They get their name from the repeated occurrence of tough beds of calcareous siltstone and fine calcareous sandstone which cap individual coarsening-upwards cycles (p.12). These hard beds form the more prominent parts of individual scars, and are riddled with trace fossils such as *Rhizocorallium* and *Thalassinoides*. Ammonites are quite common in the more argillaceous parts of the cycles, particularly the zonal form *Asteroceras obtusum*. Large body chambers of this species can be found weathering out of the shales in the rock platform (NZ 963032). *Gagaticeras* is a common indicator of the overlying *oxynotum* Zone, *Oxynoticeras* itself being rare in Yorkshire.

The Pyritous Shales mark a return to deeper water conditions, with frequent pyritic concretions and small pyritised ammonites. The best locality for these is the rock platform in the southeast corner of the bay (NZ 972025), where *Echioceras* (*raricostatum* Zone) is common. Opposite Robin Hood's Bay Town these beds, though present, are frequently covered with sand.

The full succession of the Ironstone Shales can be examined only on the north side of the bay. Note that it is very easy to get cut off on this section as the incoming tide reaches the cliff-foot very quickly at Robin Hood's Bay Town. Frequent bands of red weathering siderite nodules give the Ironstone Shales their name. The ten ammonite subzones of the Lower Pliensbachian are represented in these beds, but the ammonites are poorly preserved and sporadic. Among the genera represented are *Platypleuroceras*, *Tropidoceras*, *Acanthopleuroceras* and *Androgynoceras*. The large, semi-infaunal bivalve *Pinna* is common at some levels, flattened on the bedding planes, and current-swept belemnite accumulations occur. The Ironstone Shales become increasingly silty upwards, and grade almost imperceptably into the silty sandstones of the Staithes Formation at Castle Chamber.

A final word of caution. Robin Hood's Bay, though of great interest to stratigrapher, sedimentologist or geomorphologist, can be disappointing for the fossil collector. Most of the beds were decalcified shortly after deposition, and most of the bivalves and ammonites are preserved only as mud-filled moulds. *Gryphaea* and *Pentacrinus* are the only fossils which regularly escaped decalcification. The exposures are picked over frequently by parties. Nevertheless, the keen geologist, braving the elements in winter, may well be rewarded by interesting finds.

ITINERARY IV

Blea Wyke Point and Ravenscar

J. E. Hemingway, revised by J. K. Wright

O.S. 1:25,000 Outdoor Leisure Map Sheet 27 or
1:50,000 Sheet 94 Whitby
G.S. 1:63,360 Sheet 35 with 44 Whitby and Scalby

This is probably the most strenuous itinerary in the guide, especially as it terminates with a 150 m (500 ft) climb back up the grassy cliffs. The route is along the rock platform and over beach boulders for 2 km from Peak Steel to south of Blea Wyke Point (Figure 12). Several large rock falls and the boulder beach obscure some of the geology, and necessitate a little scrambling. The rock platform can be dangerously slippery. Blocks of shale may fall unexpectedly, and hard hats should be worn. **Low tide is necessary to study much of the section.** High tide reaches the base of the cliffs in several places, particularly at Blea Wyke Point, and care should be taken on a rising tide. Parking is available in the substantial layby at Ravenscar, overlooking Robin Hood's Bay, or alternatively southeastwards near the shop and cafe. There is a regular bus service to Ravenscar from Scarborough.

Locality 1. Ravenscar (the Peak Fault, Staithes and Whitby Mudstone Formations)

1A. Ravenscar Cliffs. From the entrance to the Raven Hall Hotel take the track to the left signposted to the beach. On reaching the cliff area, the Staithes Sandstone Formation (Upper Pliensbachian) is seen on the left-hand side of the path. Micaceous sandstones, siltstones and a thin nodular ironstone are present and yield *margaritatus* Zone fossils. On the right-hand side of the path is a gully marking the line of the Peak Fault (Figure 13). The solid exposure on the eastern side of the fault shows at its base an attenuated exposure of Yellow Sandstone (Upper Toarcian) succeeded by heavily jointed Dogger with pebble beds and by massive sandstones, siltstones and a thin coal of the Saltwick Formation. Looking back at the Middle Jurassic beds

ITINERARY IV 41

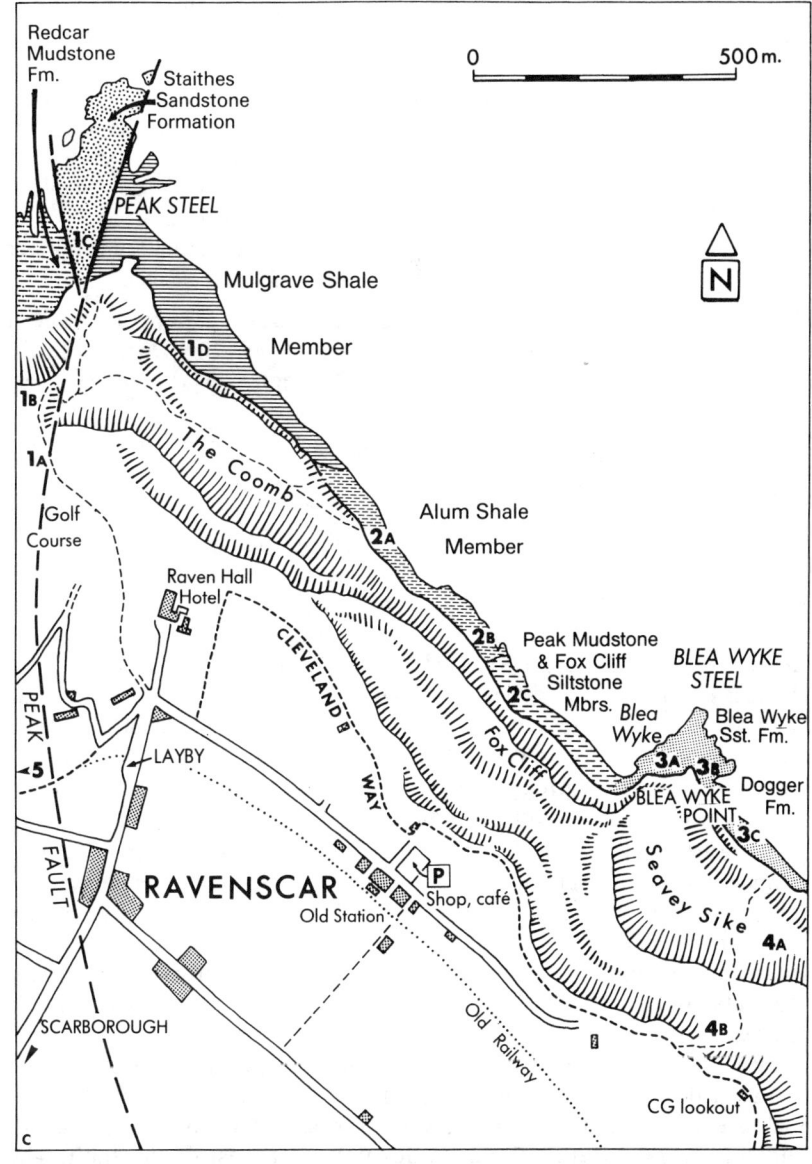

Figure 12. Map of localities in the Ravenscar area.

from slightly lower down, note that although they are on the downthrow side, the Dogger being downthrown 90 m relative to its position on the west side of the fault, the beds dip down into the fault. This suggests that the fault-plane curves at depth (i.e. that it is listric), or that there has been reversal of movement on the fault during a late compressive phase. The Peak Fault is now known to mark the western margin of the Peak Trough, a rift-like feature (p. 9-10). To the east, within the trough, a more complete Toarcian sequence is preserved than is visible anywhere else along the coast. Movement probably started during the Toarcian, and according to Alexander (1986), the fault was active, downthowing to the southeast, during Middle Jurassic times, allowing a thick sequence of fluvial sands to accumulate in the subsiding area close to the fault. Further evidence of Middle Jurassic movement of this fault will be seen later.

1B. The view of Peak Steel and Robin Hood's Bay. Pause by the fence at the cliff-edge for the classic view of the Robin Hood's Bay Dome, with the reefs of Redcar Mudstone swinging round through 180° (Figure 14). The crest of the structural dome lies seaward of the coastline within the bay, but the truncated beds on its flanks form curving scars for a distance of five kilometres to the northwest. Below, small faults branch out from the main Peak Fault, which splits into two smaller faults across the shore.

1C. Peak Steel. Follow the marked footpath to the shore arriving on the beach south of the Peak Fault, and proceed north to the fault, which is 25 m north of the ruins of the pumping station. Here, in the lower cliff, Ironstone Shales of *jamesoni* Zone age are faulted against Bituminous Shales, the upper subdivision of the Mulgrave Shale Member. This indicates a throw of 153 m. The fact that the throw of the Liassic beds is 63 m greater than that of the Dogger at the top of the cliff is a clear indication of the amount of erosion which affected the western side of the fault prior to or contemporaneous with the deposition of the Dogger (Figure 13). Much of the Upper Toarcian sequence is absent west of the fault due to this erosion.

Where the Peak Fault bifurcates on the shore, the two branches bound a prominent triangular outcrop of Staithes Formation sandstones (*margaritatus* Zone) known as Peak Steel. The 'steel', which extends seawards for nearly 500 m at low water and is covered by high tide, is stepped between Redcar Mudstone Formation to the west and Whitby Mudstone Formation to the east. The reefs are mostly heavily covered with barnacles, and the sandstone is best examined close to the cliff where this cover is minimal. Sideritic sandy limestone with profuse bivalves (*Protocardium truncatum*) alternates with fine grained, ripple-drift laminated sandstone. To the east of the fault the Jet Rock is exposed among the boulders, the Top Jet Dogger being a calcareous mudstone very different from its development west of the fault. The Bituminous Shales are well exposed southeast of the fault, and contain many crushed *Harpoceras*, *Dactylioceras* and *Pseudomytiloides dubius*.

ITINERARY IV

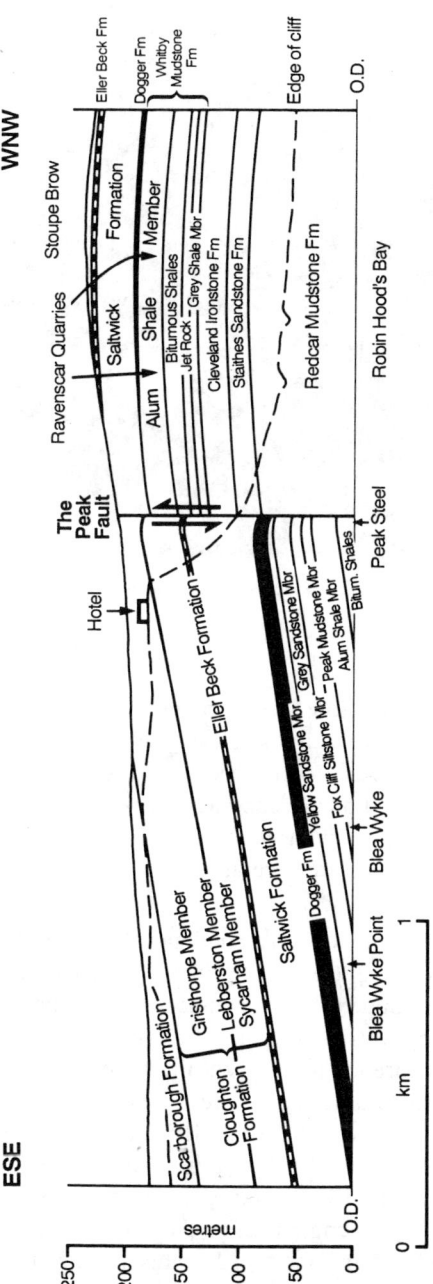

Figure 13. Diagrammatic section across the Peak Fault from near Blea Wyke Point to Stoupe Brow, as seen from the shore.

Figure 14. Robin Hood's Bay from the Peak. The curvature of the scars picks out the western flank of the Robin Hood's Bay Dome.

1D. The shore south of Peak Steel. Continue southwards for 250 m across the Bituminous Shales to reach a small bay where the Peak Stones are well seen. These are low stacks, each capped by a discoidal concretion averaging 1.2 m in diameter and showing cone-in-cone structure, a post-depositional effect. Although these are similar to the concretions in the Top Jet Dogger to the west of the fault, they are here about 7.5 m above the Top Jet Dogger level.

The upward succession is almost continuously exposed for some distance at the cliff-foot. About 500 m south of the Peak Fault, it is necessary to scramble over very slippery beach boulders to an exposure of the Ovatum Band. As at Whitby, this is a double bed of pyrite-skinned concretions associated with calcareous lenses exhibiting cone-in-cone structure. The succeeding Hard Shale unit of the Alum Shale Member is exposed for a thickness of 3 m, but the next 7 m of shale is poorly exposed due to the large spread of boulders. These lie beneath a low, grassy undercliff which sometimes shows modern lacustrine sediments on its eroded edges. It is usually possible to walk along a path on the undercliff for about 250 m until shale is again visible on the shore at Locality 2. In summer, the vegetation may force one to keep to the shore.

ITINERARY IV 45

Locality 2. The shore east of the Raven Hall Hotel (Alum Shale and Peak Mudstone Members).

2A. Northern shore exposures. The rock platform and cliff section for the next 150 m expose almost the whole thickness of the Alum Shale Member (*bifrons* Zone). The first 8.7 m belong to the *commune* Subzone. *Dactylioceras commune* is common, with *Dactylioceras* spp. and *Hildoceras sublevisoni*. The succeeding 5 m belong to the *fibulatum* Subzone. *Peronoceras fibulatum* and *Peronoceras* spp. are common, along with *Hildoceras bifrons*. The bivalve *Dacryomya ovum* is abundant, often in burrowing life position.

The 12 m of shales in the cliff, which dip down into the rock platform at the southern end of the exposure, contain several bands of cementstone concretions, and are commonly called the Cement Shales. They belong to the *crassum* Subzone and *Catacoeloceras crassum* is abundant, along with *Hildoceras bifrons* and *Pseudolioceras lythense*. The bivalve *Gresslya donaciformis* is often found in life position and *Dacryomya ovum* remains abundant. It is best to collect ammonites from the concretions scattered in the beach gravel. Large *Phylloceras heterophyllum* can usually be seen. All these shore sections at locality 2 vary in their accessibility, being sometimes covered with debris fallen from the cliff above. A more detailed description is given by Howarth (1962). For those who do not wish to proceed further along this slippery, boulder strewn shoreline, it is possible at this point to climb the low undercliff at NZ 985019 to pick up a path which winds its way up The Coomb to join the main path to Ravenscar (Figure 12).

2B. Southern shore exposures. After scrambling southwards over a large new (1991) rockfall, a further area of shale rock platform is reached. The lowest shale here was regarded as marking the base of the Peak Shales by Dean (1954). However, in a recent re-survey of these exposures (Knox, 1984), the lowest 6 m of shale at locality 2B has been transferred into the Alum Shale Member as it contains almost no silt grains. The highest Alum Shale thus extends into the *variabilis* Zone. Fragmentary *Haugia* spp., *Catacoeloceras dumortieri* and *Lytoceras* sp. can be collected. Bivalves include *Trigonia literata*, *Gresslya donaciformis* and some *Dacryoma ovum*. The overlying Peak Mudstone Member is now defined as the succeeding 12.5 m of shale containing fine silt grains, and with scattered concretions. Interesting ammonites include *Haugia* and *Denkmannia*. *Trigonia literata* is common but *Gresslya* and *Dacryomya* no longer occur.

2C. Fox Cliff. The highest 4.5 m of the Peak Mudstone Member is exposed in the base of Fox Cliff. It was formerly included in the Striatulum Shales, a division now abandoned. The siderite nodules from this 4.5 m of shale are very fossiliferous with abundant *Grammoceras* sp. and a large variety of bivalves, including *Camptonectes* sp., *Ostrea* sp., *Pteria substriata*, *Protocardium* sp., *Pecten* sp. and *Inoceramus* sp. The succeeding 11 m of beds exposed in Fox Cliff contain medium and coarse silt grains and have been named the Fox Cliff Siltstone Member of the Whitby Mudstone Formation by Knox (1984). This member is difficult of access in the cliff, but fallen sideritic concretions are common, yielding *Pseudogrammoceras* sp. (*thouarsense* Zone) in addition to the bivalves mentioned above.

Proceed now over the poorly-exposed area of boulder-strewn beach into the centre of Blea Wyke. There is a magnificent section of Ravenscar Group sediments in the cliff. Notice in the beach the blocks of strikingly buff, fine grained sandstone showing slump structures, which have fallen from a horizon about 6 m below the base of the Scarborough Formation near the cliff top. Thin coals may readily be made out in the cliff.

Locality 3. Blea Wyke Point (Blea Wyke and Dogger Formations)

3A. The northern side of Blea Wyke Point. In the north face of the Point is the 9 m thick type-section of the Grey Sandstone Member of the Blea Wyke Formation. These high Toarcian sandstones and siltstones here attain their maximum development in Yorkshire. The base is marked by a change from the Fox Cliff Siltstone to paler, burrow-mottled sediment, with coarse silt- and fine sand-filled burrows. The dense, smooth carbonate concretions of the Fox Cliff Siltstone are no longer present. The subzonal ammonite *Phlyseogrammoceras dispansum* occurs in sandy concretions particularly in the lower beds within Blea Wyke. *Lingula beani* is present throughout and *Orbiculoidea reflexa*, *Dentalium elongatum* and *Homeorhynchia cynocephala* also occur. Belemnites are very common. The two massive sandstones totalling 3.6 m in thickness, which cap the Grey Sandstone and form the wave-cut platform around the base of Blea Wyke Point, are known as the Serpula Beds (Rastall & Hemingway, 1939). They contain masses of the tubes of *Serpula deplexa* up to 30 cm across, which weather into burrs on the sandstone surface. The Serpula Beds are also characterised by an abundance of other trace fossils which invest the rock almost completely. Each of the massive sandstones is most fossiliferous in its topmost few centimetres, with *Phlyseogrammoceras* sp., *Homeorhynchia cynocephala*, *Pteria* sp., *Oxytoma* sp., and *Orbiculoidea reflexa* (Dean, 1954).

3B. The southern side of Blea Wyke Point. The upper Yellow Sandstone Member of the Blea Wyke Formation is exposed in the cliffs just south of Blea Wyke Point. These beds yield only a poorly-preserved fauna, but careful collecting has produced *Dumortiera* spp., *Homeorhynchia cynocephala*, 'Terebratula', aff. *trilineata*, *Gresslya* sp., *Modiolus* sp., *Pteria* sp., *Pinna cuneata* and *Trigonia* sp. (Dean, 1954).

3C. South of Blea Wyke Point. Continue southeast up the succession and across boulders to the Terebratula Bed at the base of the Dogger Formation. This grey-hearted, sideritic sandstone (0.45 m) is rich in *'Terebratula' trilineata*, with less common *Gresslya donaciformis*, *Trigonia ramsayi*, *Pentacrinus* sp. and fossil wood. The bed is usually firmly cemented to the delicately bored surface of the Yellow Sandstone. The time gap in between is considered to be small, of the order of one subzone (Parsons, 1980). Three metres below the top of the Dogger Formation is the Nerinea Bed, a lenticular shell-bed up to 30 cm thick rich in *Nerinea cingenda*, *Cerithium* sp., *Astarte* sp., *Trigonia* sp., *Gervillia* sp., etc. The original calcareous shells are replaced by siderite and weather into sharp relief from a matrix of sandy chamosite oolite. **This unique exposure, because of its excellence, should not be hammered**, but fossil collections can be most readily made from fallen blocks. Southwards, the surface of the Dogger Formation, here a sphaerosiderite passing down into a sandy, sideritic chamosite oolite, is deeply weathered, with prominent ribs marking the joints.

Locality 4. The Cliffs south of Blea Wyke (Ravenscar Group).

South of Blea Wyke Point the low cliff ends, and it is possible to ascend via a narrow path leading across the undercliff. The way up begins at the point where the Dogger Formation disappears below sea-level at high tide. There is no 'made up' path here, but simply a route tramped down by fishermen and geologists over the years. In late summer the path is rather obscured by bracken and care must be taken to avoid falling over loose boulders. The Eller Beck Formation makes a well marked feature 50 m above the Dogger Formation.

4A. Cliffs 600 m south of the path. The attenuated Lebberston Member is not exposed near the path but may be seen 600 m to the south about 27 m below the Scarborough Formation. The Lebberston Member here is essentially a ferruginous marine sandstone 3 m thick with siderite mudstones containing *Trigonia* sp., *Ostrea* sp. and *Haploecia straminae*.

4B. The upper cliff. Ascend steeply via the overgrown tip from the old workings for the limestone of the Scarborough Formation. The path now runs up to the right below crags of sandstone and limestone. A thin coal with seatearth may be examined about 3 m below the base of the Scarborough Formation. Such coals, dominated by *Equisetum*, are frequently found at this level, but rarely exceed 0.5 m in thickness. The Scarborough Formation attains its maximum thickness of 35 m at Ravenscar, and is divisible into four main units as follows (after Parsons, 1977; Gowland & Riding, 1991):–

  m

5. Bogmire Gill Member:– fine-grained, flaggy sandstone with
   abundant moulds of *Pleuromya* sp.   4.5

4. Ravenscar Shales:– sandy shales with concretions rich in
   Pseudomonotonis lycetti, Perna sp., *Ostrea* sp. and belemnites. 17.0

3. Hundale Sandstone and Spindle Thorn Limestone:–
   calcareous sandstones and impure limestones with *P. lycetti* and
   Gervillia scarburgensis.   5.5

2. Hundale Shale:– fine grained, flaggy sandstone with *Pleuromya* sp. 4.5

1. Helwath Beck Member:– thick, laminated sandstone showing
   cut-and-fill structures with frequent marine trace fossils
   including *Diplocraterion*.   seen to 3.0

Unit 1 is now included in the Scarborough Formation (Gowland & Riding, 1991). Previously, it had been considered to be the highest unit of the Gristhorpe Formation (Parsons, 1977). Two massive beds of limestone from unit 3 stand out. Most of the fossils, which include *Lopha, Astarte, Pecten, Modiolus, Teloceras, Pentacrinus* and belemnites, are weathered and very fragile. The Ravenscar Shales and Bogmire Gill Member are now largely grass covered and cannot be examined without an excavation. Continue to the cliff-top and return to Ravenscar.

Locality 5. Ravenscar Brickworks (Fault-attenuated Whitby Mudstone Formation and Ravenscar Group successions).

The succession between Peak Steel and Blea Wyke Point, on the east side of the Peak Fault, should be contrasted with that in Ravenscar Quarries 450 m west of the fault (NZ 973015). From the entrance to the Raven Hall Hotel proceed down the track heading southwestwards (Cleveland Way). Carry on the main track until an archway on the left is reached under the abandoned railway. A broad track leads through the archway into the old brick-pit. The area around the kilns has been fenced off as they have become dangerous. The high, inaccessible face of the quarry shows a thin (1 m maximum) development of the Dogger Formation resting on Alum Shales, the former in turn overlain by level-bedded alternations of shale and sandstone of the Saltwick Formation. Many blocks of the Dogger have fallen to the quarry floor. Some contain bored pebbles of derived Liassic nodules. 'U' shaped *Diplocraterion* burrows, some superbly developed, penetrate the full thickness of the Dogger Formation from its top surface, being filled with coarse quartz sandstone. Many blocks of plant-rootlet sandstone have fallen from the Saltwick Formation, and from here Sargeant (1970, pl.21) figured a dinosaur footprint, referred to *Satapliosaurus*. Southeast of the quarry, close to the Peak Fault, the Dogger Formation thins out entirely. Some 57 to 58 m of beds, including also the Cement Shales, Peak Mudstone, Fox Cliff Siltstone and the Blea Wyke Formation, are absent on the western, upthrown, side of the fault. The younger beds also show a marked reduction in thickness from east to west across the fault, the Saltwick Formation being reduced from at least 57 m to only 30 m and the Cloughton Formation from 77 m to about 55 m. The differences in the successions on the two sides of the Peak Fault are shown in the cross-section (Figure 13).

The omission of nearly 50 m of Liassic beds, and the attenuation of the Ravenscar Group by some 60 m, have been explained in two ways. Fox-Strangways and Barrow (1915) suggested that contemporary Jurassic movement on the Peak Fault allowed Upper Toarcian beds to be preserved on the downthrown side, while on the shallow upthrown side they were either scoured off or not deposited. Additional mid-Jurassic movement is necessary to account for the difference in thickness of the Middle Jurassic beds. In contrast, Hemingway (1963, 1974) and Hemingway and Ridler (1982) suggested that the successions had been laid down in areas which were initially some distance apart, and which had been brought together by transcurrent (lateral) movement along the Peak Fault, probably during the Tertiary. Sinistral movement would be necessary to account for the distribution of the beds.

A structural analysis of the Peak Fault Zone as it extends south to Cayton Bay shows that it can be interpreted as a transcurrent fault system, but one of dextral movement, not sinistral. The main northwesterly trending faults and the subsidiary NNE/SSW trending branch faults fit naturally a near north-south primary stress pattern. ESE/WNW trending secondary drag folds fitting this stress pattern can be recognised at Osgodby Nab (Wright, 1968, fig. 9) and at Hayburn Wyke. Limited dextral movement of the Peak Fault Zone is very likely, but the major sinistral movement required by Hemingway's hypothesis is now considered unlikely.

Milsom and Rawson (1989) have traced the Peak Fault Zone into the offshore area, and define a Peak Trough, a zone of trough-faulting 5 km wide bounded to the west by the Peak Fault, and to the east by a series of faults which will be seen in Itinerary VII running from Scarborough to Red Cliff. Normal faulting took place in the trough in Triassic, mid-Jurassic, probably Late Jurassic/Cretaceous and Tertiary times. Movement was by gentle creep rather than sudden earthquake shock for much of the time. The surface expression of the trough may have been only a metre or two (Alexander, 1986) and as such the trough operated at Ravenscar for much of the Aalenian and Bajocian, allowing a succession to accumulate within the trough 60 m thicker than that on the western flank. The Peak Fault Zone now takes its place amongst the many other Mesozoic synsedimentary fault systems which have been discovered recently in the North Sea and in southern England.

ITINERARY V

Cloughton Wyke to Scalby Ness

J. K. Wright

O.S. 1:25,000 Outdoor Leisure Map Sheet 27 or
1:50,000 Sheet 101 Scarborough
G.S. 1:63,360 Sheet 35 with 44 Whitby and Scalby

This itinerary provides a traverse through part of the Ravenscar Group, from the Lebberston Member (Millepore Bed) to the Scalby Formation (see Table II). The full excursion involves a 5 km walk along the rock platform at the base of the cliffs and over stretches of beach boulders. The going can be quite strenuous. However, it is possible to avoid much of the more strenuous walking by ascending the cliff at convenient points and proceeding via the cliff-top path. **A falling tide is necessary** to see the geology to its best advantage, particularly as the full sweep of the meander belt sandstone in Scalby Bay (Locality 4c) can only be seen within two hours of low tide. Other parts of the section are not accessible above mid-tide. Hard hats are essential. It is possible to drive cars or minibuses down to the cliff top from both Cloughton and Burniston (Figure 15), though the car parks are small (half a dozen vehicles) and the roads narrow, so that large parties will have to be dropped off in Cloughton village at the start of the itinerary. There is a regular bus service from Scarborough to Cloughton.

Locality 1. Cloughton Wyke (Lebberston Member to basal Scarborough Formation).

1A. The north side of Cloughton Wyke. Descend to the shore from the car park at Cloughton Wyke and walk northwards over large fallen sandstone blocks (some of which show excellent sedimentary structures) for 250 m. Here the Lebberston Member, represented by the Millepore Bed, forms a series of reefs running southeastwards out to sea. It consists of three tiers of shelly, calcareous sandstone and sandy limestone each forming a coarsening-upwards cycle. The middle tier is the most fossiliferous and the upper part is extremely hard where it is cemented by siderite. The fauna includes *Arcomya elongata, Lima duplicata, Entolium demissum, Pleuromya beani, Trigonia* sp., *Pholadomya saemanni* and *Modiolus imbricatus*. The branching bryozoan *Haploecia [Millepora] straminea*, which originally gave its name to this unit, is less common here than at Osgodby Nab (Itinerary VII). Many of the shells are preserved in the white-weathering clay mineral dickite. Excellent cross-bedding makes the upper tier very distinctive. Compared with the sections south of Scarborough (Itinerary VII) the Millepore Bed here

ITINERARY V

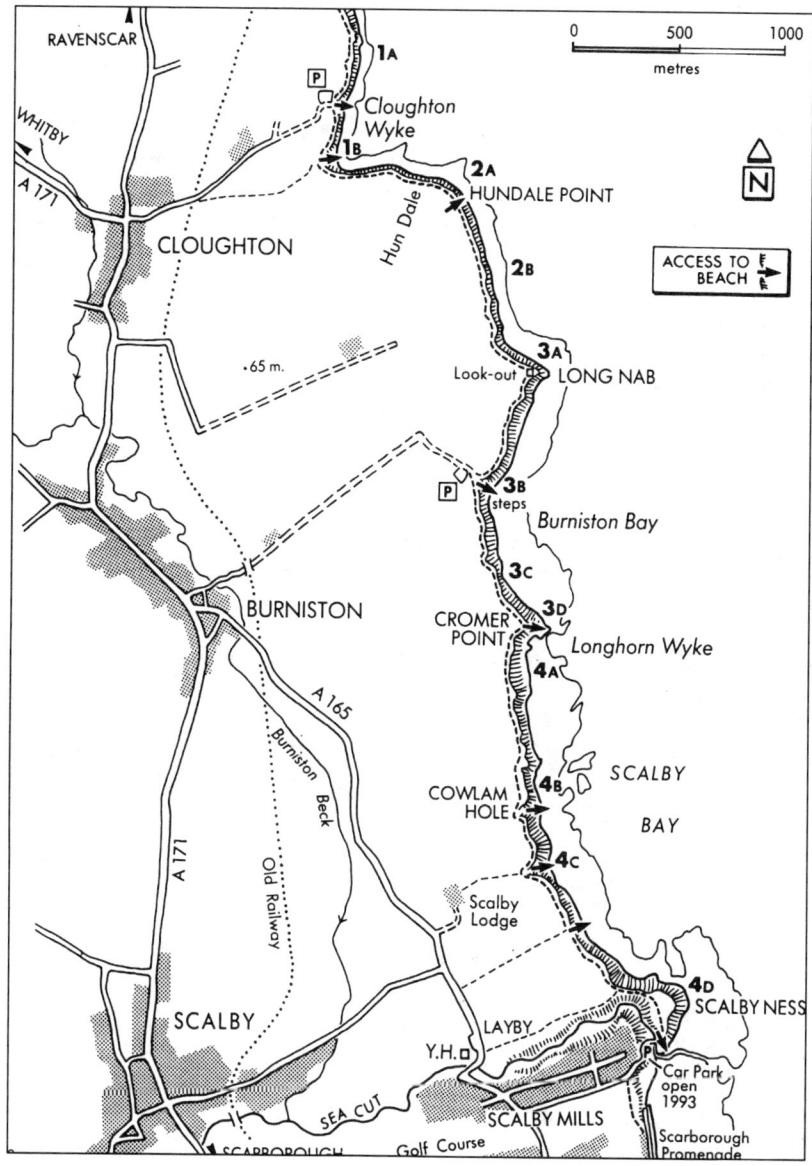

Figure 15. Map of localities between Cloughton and Scalby.

shows a marked decrease in the content of ooliths and shell debris and an increase in calcareous sandstone and particularly in iron carbonate, indicating that it accumulated nearer the shoreline (Bate, 1959).

Following a suggestion of Knox (personal communication) the succeeding Yons Nab Beds are regarded here as the basal, quasi-marine part of the Gristhorpe Member. The Yons Nab Beds are exposed in the low cliff on the way back towards Cloughton Wyke. There they interdigitate with fully non-marine beds. Resting on the Millepore Bed is 0.6 m of ripple-drift cross-laminated sandstone, succeeded by 1.6 m of flaser-bedded, intertidal, shaly sandstone with numerous bivalves including *Trigonia, Pecten* and *Ostrea*. Variably bioturbated, ripple-drift cross-laminated sandstone (0.6 m) seems to complete the Yons Nab Beds sequence. It is succeeded by 1 m of shale with coaly, carbonaceous layers, which weathers back readily.

However, traversing southwards across a large rock-fall, a low cliff is reached exposing a further 2.6 m of flaggy and well-bedded, quasi-marine sandstone containing occasional *Diplocraterion* burrows, and strongly bioturbated towards the base. These beds were first included within the Yons Nab sequence by Livera and Leeder (1981). The top of this unit forms a slippery ledge facing southwards and above it occurs a fully non-marine succession of 4.5 m of shales with thin coals and a rootlet bed, overlain by thick sandstone.

1B. Cloughton Wyke. The non-marine sandstones and shales of the Gristhorpe Member are exposed in the cliffs and rock platform for the 0.5 km from the car park area southwards to the centre of Cloughton Wyke. The lower beds, coming above the sequence described at Locality 1A, are exposed in the cliffs on either side of the boulder-strewn beach below the car park. They consist of thick, planar-bedded sandstones, a facies unusual in the non-marine beds, and which is almost certainly the result of sheet flooding. This can be demonstrated by a close examination of these beds in the cliff between the car park and Cloughton Wyke. **The cliff is, however, unstable with considerable overhangs, and great care should be taken along this section.**

The planar-bedded sandstones generally occur in four half-metre tiers. Below the lowest tier there is a gradual shallowing-upwards sequence extending over a thickness of 1.5 m from shale with siderite nodule bands through silty shale into an excellent, thin-bedded, ripple marked sandstone. The basal tier of massive sandstone infills grooves and flutes carved into the thin-bedded sandstone. The bottom structures indicate a current direction flowing from NNE to SSW. The thick beds of sandstone thus appear to be associated with a large distributary channel to the northeast, which was subject to repeated

Figure 16. An *Equisetites* rootlet bed in the Gristhorpe Member, Cloughton Wyke.

breaches such that extensive sheet sands (crevasse-splays) spread southwestwards into the Cloughton Wyke area. After each episode of crevasse-splaying, the low lying area around Cloughton Wyke was then colonised by marsh plants. Rootlets are clearly visible in tiers 2 and 4 (Figure 16). The highest tier is overlain by a thin coal. Within each tier, though the bedding is frequently planar reflecting the strong current flow, ripples and cut-and-fill structures are also seen. In Cloughton Wyke itself some 7 m of thin sandstones, siltstones and shales with rootlet beds overlie the crevasse-splay sandstones.

These are followed, 4 m above high water mark and adjacent to the steps up the cliff, by a 1 m bed of intensely balled-up sandstone. The convolution is thought to have been caused by an earthquake shock due to movement along the Peak Fault, here running 1.6 km to the west. The bed can be traced all the way to Ravenscar. Some 1 m below the convoluted bed, marine borings descend into the underlying pale grey siltstones. Gowland and Riding (1991) have thus grouped the convoluted sandstone, the overlying 2 m of silts and shales and the succeeding 4.2 m bed of sandstone with numerous *Diplocraterion* burrows into the Helwath Beck Member at the base of the Scarborough Formation. A varied trace fossil assemblage is present, with very rare marine bivalves (Gowland & Riding, 1991).

Amid the boulders on the shore (TA 02109495), the dark grey, micaceous, silty clays of the highest Cloughton Formation yield an abundant flora, usually carbonised, but occasionally preserved in a flexible, chitinous form. The flora is rich, but individual species have a remarkable local distribution. They include *Ptilophyllum pecten*, *Cladophebis* sp., *Czechanowskia* sp., *Nilssoniopteris* sp., and *Otozamites* sp. (J.E.H.).

Locality 2. Hundale Point (Scarborough and Scalby Formations).

If the tide is in, it may be advisable not to attempt the scramble over the large sandstone blocks close to the cliff. It is possible to ascend via steps and a step-ladder in the centre of Cloughton Wyke and walk along the cliff-top path to the easy descent at Hundale Point. **Do not attempt to decend to the beach at Hundale as there is a sheer drop.**

2A. Hundale Point. Between Hundale Scar and Hundale Point (Figure 15) the full sequence of the Scarborough Formation is well displayed in the rock platform and base of the cliff. The following is adapted from Parsons (1977) and Gowland and Riding (1991):–

	m
7. Bogmire Gill Member:– micaceous, sandy siltstone passing up into fine grained sandstone.	approx. 2.5
6. White Nab Ironstone:– sulphurous, grey, sandy shales with three layers of iron-rich concretions.	1.3
5. Ravenscar Shale:– dark grey shales with small, fossiliferous concretions.	8.2
4. Spindle Thorne Limestone:– grey, sandy, calcareous mudstones with a hard, grey, sandy limestone (0.2 m) at the top.	3.7
3. Hundale Sandstone:– two coarsening-upwards units passing from siltstone into calcareous sandstone, and separated by a thin, argillaceous limestone.	4.0
2. Hundale Shale:– silty sandstone passing up into muddy limestone.	2.6
1. Helwath Beck Member:– Massive, coarsening-upwards, delicately cross-bedded sandstone with abundant marine trace fossils, particularly near the base and in the top surface.	seen to 3.5

Diplocraterion is very well seen in many of the fallen blocks of Helwath Beck Member along the E-W stretch of coastline in the Wyke, and in the reef running out into Hundale Scar. Above, the tough, calcareous siltstones, sandstones and thin, muddy limestones of the Hundale Shale and Hundale

Sandstone Members form a series of resistent beds running across the rock platform and forming the feature of Hundale Point. The Hundale Sandstone was formerly referred to as the Crinoid Grit (Parsons, 1977). A diverse bivalve and gastropod fauna is largely restricted to the thin limestone/shale beds. *Gervillia scarburgense* is abundant in the Hundale Shale sandstone and the trace fossils of the Hundale Sandstone are worth close inspection (Farrow, 1966; Romano & White 1987; Gowland & Riding, 1991).

The Spindle Thorne Limestone comprises grey, sandy, calcareous mudstones and siltstones exposed in the base of the cliff near the mass of fallen blocks of Moor Grit, which cover the rock platform south of the point. A varied bivalve fauna occurs, with occasional ammonites towards the top of the member. The three shaly highest members of the Scarborough Formation are exposed in the cliff section by the path descending to the shore. Occasional small, round concretions in the Ravenscar Shale contain uncrushed fossils including ammonites, particularly *Dorsetensia* and rarer *Stephanoceras*. The bivalve fauna is very restricted (Gowland & Riding, 1991). The nodular siderite at the top of the White Nab Ironstone yields many bivalves, gastropods and occasional ammonites.

2B. The cliffs south of Hundale Point. The Scarborough Formation passes rapidly beneath beach level southwards, and the cliffs and rock platform all the way to Scarborough are composed of Scalby Formation sandstones and shales. The sandstones of the Moor Grit Member form the cliff along much of this unnamed bay south of Hundale Point. The 10 m of Moor Grit shows giant sets of trough cross-stratification indicating an east to northeast current flow. A drawing of this cliff section appears in Nami and Leeder (1978). The coarse, mature sandstones are rich in coalified plant remains with subordinate charcoal fragments. The massive channel sands alternate with and pass laterally into off-bank laminated siltstones and crevasse-splay sheet sands which alternate with shale. The impression is of a rapidly accumulating series of beds, with little re-working or lateral migration of channels.

Locality 3. Burniston Bay (Scalby Formation).
3A. Long Nab. Proceed round the bay to Long Nab, which is protected from erosion by several huge blocks of cross-bedded sandstone up to 4 m thick derived from a channel in the lower part of the Long Nab Member. The cliff section from the south side of Long Nab into Burniston Bay displays well the main features of the higher part of the member. There are alternations of fine clays and siltstones with occasional sand-filled channels. A small channel is

seen in section at eye-level, cutting down half a metre into the underlying shale. The sandstones above are more persistent and even-bedded. At a height of 4 m above the rock platform occurs the well-known Burniston Footprint Bed. The dinosaur prints were made in soft, silty clay subsequently infilled by a gentle incursion of silt and sand, preserving the prints as casts. As blocks of this silty sandstone fall to the beach they frequently come to rest upside-down at the foot of the cliff, displaying the footprints very well (Figure 17). Two types were figured by Black *et al* (1934).

The rock platform throughout the bay reveals a complex series of intersecting channel sandstones of the meander belt unit (Nami, 1976). Cross-bedding dips of 20° or 25° are common. The sands accumulated as point bar deposits on the inside bends of migrating channels. Thus, the strike of the (epsilon) cross-bedding indicates the orientation of the channel, with the dip of the cross-bedding pointing down into the channel. Most of the cross-bedded units are curved in plan, showing the meandering nature of the stream complex, with later channels cutting down into and intersecting the bedding of earlier channels.

3B. Burniston Steps. As one reaches the low cliffs on either side of Burniston Steps one comes repeatedly to sections showing such similar features that a single sketch (Figure 18) can be used to illustrate them. Above the meander belt sandstones come 2 to 3 m of grey, alluvial clay with subordinate silty sandstones. One, and frequently two beds of dark, carbonaceous clay are present full of flattened plant stems. Rootlets pass down into the meander belt sandstone. Sideritisation in the clay and upper meander belt sandstone is intense, both as large concretions and as beds of sphaerosiderite. These ironstones are interpreted as evidence for the existence of ancient soil profiles (the palaeosols of Kantorowicz, 1990). Siderite in such quantities would be expected to precipitate from oxygen deficient ground water beneath peaty soils, represented here by the carbonaceous clay.

Thin alternations of sheet sandstone and silty clay follow, with the development of at least a dozen localised channels many of which were first described by Black (1928), and are known by Black's letters A to H. In almost all sections one can demonstrate that level-bedded alternations of siltstone and sandstone pass laterally into strongly cross-bedded channel sandstones, with the channels cutting through the clays and frequently resting on meander belt sandstone. The Footprint Bed is just above the horizon at which channels develop, and is succeeded locally by a thin marine sandstone. Close to Burniston Steps the footprint sandstone forms an overhanging ledge. **Please do not hammer here.**

Figure 17 (*Upper*). Saurian footprint in a fallen block from the Burniston Footprint Bed, Long Nab. *(Lower)*. Point-bar sands infilling a channel in the meander belt in Scalby Bay. Stream direction from left to right.

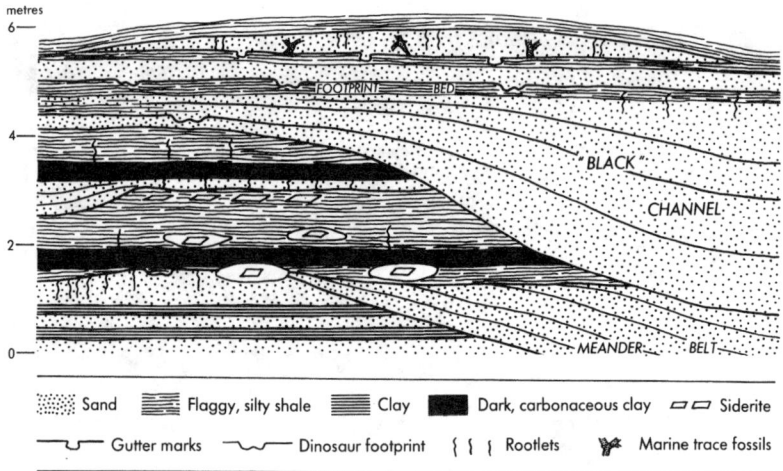

Figure 18. Diagrammatic section of the Lower Long Member, Burniston and Scalby Bays.

3C. The south end of Burniston Bay. In the southeastern corner of Burniston Bay a, small channel is seen just above the beach. Above comes the Footprint Bed, and then a bed of sandstone increasing to 90 cm in thickness southeastwards. Many large blocks of this sandstone have fallen onto the shore. The bed consists of a well sorted, ripple-drift laminated, fine grained sandstone. When blocks have turned over in falling, the lower surface reveals numerous infilled burrows of the marine trace fossil *Ophiomorpha* (Livera & Leeder, 1981). The marine origin of this bed seems certain. However, the top surface of the bed, seen in the cliffs and in blocks that have not rotated, contains numerous rootlets. So the marine episode was brief. Near the northern end of this exposure, the marine bed is scoured by a later channel and locally replaced by cross-bedded sandstone. At the southeastern end of the exposure, where the marine bed is wedging out, its underside reveals spectacular infilled scour channels or gutter marks running parallel to the cliff. The slightly sinuous channels are at least 7 m long, 10 cm deep and 'U' shaped. Upturned blocks on the beach show level-bedded sandstone infilling the channels. This marine bed occupies no more than 120 m of the cliff section at this end of Burniston Bay and appears to be a lens almost certainly infilling a shallow channel scoured by a tidal surge. Thin, localised deposits of a similar nature occur elsewhere in Burniston and Scalby Bays at the same horizon.

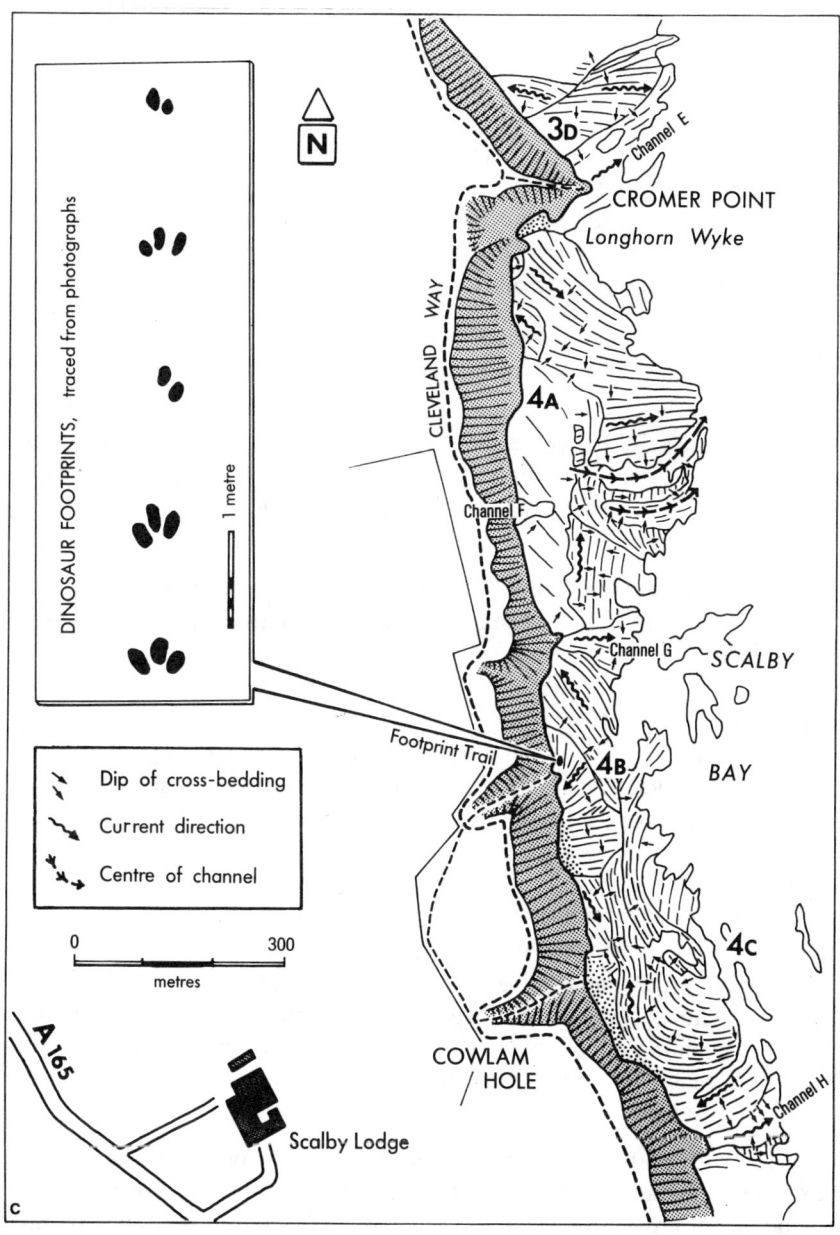

Figure 19. Map of meander belt channels in Scalby Bay (based on field mapping by JKW, 1969).

3D. Cromer Point. At Cromer Point, Black's Channel E is seen well, there being a rapid lateral transition from off-bank laminated siltstones and sandstones into massive channel sandstone through which the present sea has cut a natural arch.

Having rounded Cromer Point it is necessary to negotiate Longhorne Wyke. The cliff on the south side of this gully falls sheer into the sea for much of the tidal cycle, and **it is not possible to get round it within 3 hours either side of high tide.** One can ascend the cliff by the rough path running up the north face of the headland, or make one's way up the grassy cliff in the centre of Longhorn Wyke, rejoining the excursion at Locality 4B.

Locality 4. Scalby Bay (Scalby Formation).

South of Cromer Point, within the long stretch of Scalby Bay, there is a sucession of small bays cut in the soft siltstones and clays of the Long Nab Member, with promontaries in between formed by Black's sand-filled channels projecting from the cliff (Figure 19). **Cliff falls take place regularly here, and care should be taken.**

4A. The beach and cliff south of Longhorne Wyke. As one proceeds south from Cromer Point cross-bedded meander belt sandstones are excellently displayed in the rock platform (Figure 17). Small-scale cross-bedding within the massive, southerly dipping beds, particularly the infilling of small chute channels by trough cross-bedding, shows that the current flow was eastwards.

On reaching the large pile of sandstone blocks marking the next channel (TA 029925), two small sandy bays are seen, with channels F and G forming promontaries. For 100 m south of the sandstone blocks, the cliff section reveals a whole series of small, sand-filled channels stacked one above the other, some choked with carbonised remains of tree trunks, and with much slumping. This complex series of beds passes laterally into sheet sandstones and shales with occasional dinosaur footprints.

Out on the rock platform the position of the meander belt is occupied by a series of almost level-bedded, ripple-marked sandstones alternating with shales containing plants. Here, we have the original sediments laid down above the Moor Grit, which through most of the area have been reworked to form the meander belt.

4B. The rock platform northeast of Scalby Lodge. South of Channel G, an excellent meander belt channel is seen in the rock platform, with a predominant cross-bedding dip to the east, and a subsidiary westerly dip. Infilled chutes show a northerly current flow. This channel cuts across a series of almost flat lying sandbanks which gradually acquire a southerly dip

ITINERARY V 61

southwards. At the foot of the small valley, the sandbank nearest present highwater mark displays a 10 m long trail of dinosaur footprints (Figure 19). Photographs of this locality were published by Delair and Sargeant (1985, fig. 3). The poorly sorted fluvial sand was compacted under and around the prints, which are now being revealed by the selective erosion of the overlying argillaceous sandstone.

4C. Cowlam Hole. Opposite this southernmost of the two small valleys (Figure 19), a meander belt channel makes a spectacular 180° sweep across the rock platform. Cross-bedding dips swing round from northeast through southwest to southeast. Trough cross-bedding within the chutes shows that the current flow was clockwise. Seawards, a later channel cuts across the meander and also cuts across several earlier channels to the north.

4D. Scalby Ness. Proceed now across Black's Channel H into the southern end of Scalby Bay and round onto Scalby Ness (Figure 15). The gentle seaward dip carries Long Nab Member shales which overly the meander belt into the base of the cliff. These silty shales, the Scalby Plant Bed, contain frequent well-preserved leaves of *Ginkgo huttoni*.

ITINERARY VI

Egton Bridge and Goathland

J. E. Hemingway

O.S. 1:25,000 Outdoor Leisure Map Sheet 27 or
1:50,000 Sheet 94 Whitby
G.S. 1:63,360 Sheet 43 Egton

This short excursion is designed to occupy a morning or an afternoon when tidal conditions preclude shore geology. It comprises a car or minibus tour over the southern side of the Esk Valley from Egton Bridge up to Goathland (Figure 20). The itinerary is not suitable for coaches. Much of it is concerned with the effects of the last (Devensian) glaciation upon the topography and drainage pattern of the area, although visits are also made to two rock units not seen in the coastal itineraries. Walking is minimal, with the exception of Locality 8, where it is necessary to ford Eller Beck.

The region around the villages of Egton Bridge and Goathland displays most clearly the glacial phenomena consequent upon the retreat of the ice-front from Cleveland to the sea. During Devensian times ice impinged on N. E. Yorkshire from the west, north and east. To the west, the Vale of York ice pushed south below the western escarpment, but failed to over-ride it effectively. From the northeast more vigorous ice drove inland from the North Sea over the coastline for a distance of up to 10 km. The middle of N. E. Yorkshire remained ice-free (Figure 5), though undoubtedly snow-covered, and its drainage valleys were effectively ice-blocked. Meltwater produced during glacial retreat therefore accumulated in the valley heads until it overflowed across confining spurs, usually (but not always) at their contact with the ice-front. The recognition of this phenomenon in N. E. Yorkshire by P. F. Kendall (1902) resulted in this region now being regarded as classic in the demonstration of marginal glacial drainage channels, glacial lakes and their related features and deposits.

The Egton Bridge-Goathland area is essentially a broad spur between the east-west valley of the Esk to the north and the north-south valley of the Murk Esk and its parent tributaries of Eller Beck and West Beck to the east. During the ice-retreat waters from the 17 km long glacial Lake Eskdale to the west drained round the edge of the spur and eroded it into a series of marginal channels along the ice-edge at successively lower levels. The floors of these channels are now hidden under peat, but their upper slopes are virtually unaltered in the sharpness of their incision. The itinerary outlined here aims not only to demonstrate Kendall's interpretation of these features but also to point out such parts of the solid geology as merit attention.

Locality 1. Duckscar Quarry, Egton Bridge (Cleveland Dyke).
There are no problems of access to this disused roadside quarry at NZ 798053, and parking is available for several cars. The quarry face reveals a well-exposed section in a Tertiary tholeiite, the Cleveland Dyke, here 9 m wide. Tholeiite is a type of basalt, usually lacking in olivine and non-porphyritic. The Whitby Formation is the country rock, yields abundant *Pseudomytiloides dubius*, and is most probably at the level of the Bituminous Shales. Metamorphism is slight, but a few centimetres of baked and whitened shale may be seen welded on to the outstanding weathered walls of the dyke.

Locality 2. Moss Swang. Return to Egton Bridge, and follow the road signposted Goathland. At the crossroads (NZ 807038) stop and observe the great, flat-floored channel of Moss Swang just west of the road. This is one of the finest of the marginal channels which carried water draining south from Egton Grange Lake. Here, both sides of the channel are cut in solid rock. Turning to the east, observe the view down lower Eskdale and across the Eskdale Dome. This structure yielded natural gas from which Whitby and district was supplied in the 1960's. Notice in particular the flat-topped feature at approximately 76 m O.D. which extends along the valley side near Newbeggin Hall (NZ 841069) for nearly 2 km. This surface, cut in boulder clay, may be traced at rather higher levels above Grosmont and is a high terrace of the late-glacial Esk graded to the 58 m sea-level.

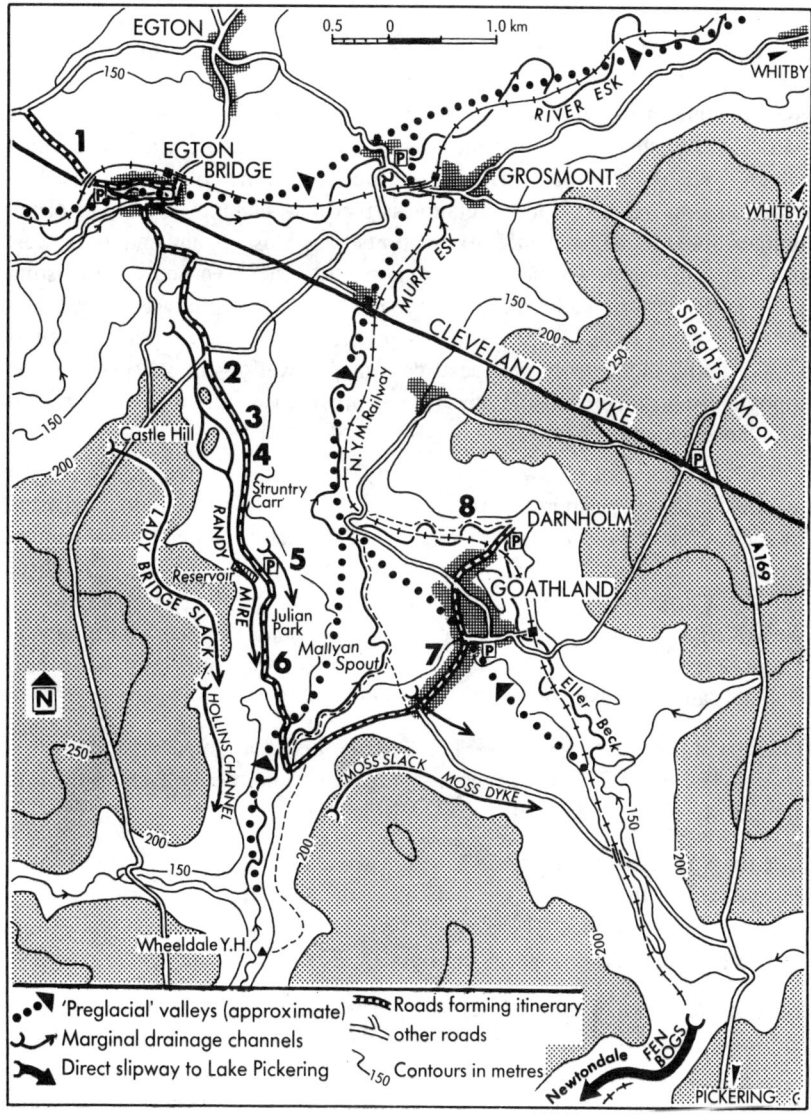

Figure 20. Glacial and pre-glacial drainage in the Egton Bridge – Goathland area, showing reversal of drainage directions. Except in the river gorges, drift covers all the area below 180 m O.D.

Locality 3. Castle Hill. At NZ 810032 park and observe the unusual pattern of channels surrounding the outlier of Castle Hill. West of Castle Hill a channel cuts about 15 m into solid rock, but hangs at its northern end in particular 15 m above the main Moss Swang channel. Kendall regarded this as indicative of the readvance of a small lobe of the ice front over the older Moss Swang channel, so that meltwaters were forced to cut a new channel which reached a depth of only 15 m before a recurrence of retreat caused its abandonment. A second interpretation is that both channels were part of the uninterrupted retreat sequence: excavation along Moss Swang and to the west of Castle Hill cut down for 15 m; a minor retreat then exposed a pre-existing gully in the eastern wall which, because of its greater depth, was followed by the melt-waters and deepened to its present form. This latter opinion is supported by the absence of glacial deposits, as well as the sharpness of the topography on the north side of Castle Hill, which would not be expected if readvance had taken place.

The possibility of meltwaters draining through and particularly below the ice must also be considered (Gregory, 1962). Many channels undoubtedly cut by meltwaters are clearly not marginal to an ice sheet. Modern, near-stagnant ice shows that much, if not the majority, of meltwater passes through or below the ice, initially by fissures and crevasses, and not marginally round the ice edge. Many topographic features, inexplicable as marginal drainage phenomena, are clearly formed by these processes and the Castle Hill channels are most likely to be of this origin.

Locality 4. Struntry Carr (NZ 811026). In this area the channel as such is absent. The moor edge escarpment is sharpened and the meltwaters flowing along it were contained on their eastern side by the ice wall.

Locality 5. Randay Mere (NZ 811019). A car park is available on the eastern side of the road. Here, the meltwaters have cut a fine channel through solid rock on both sides. The channel is now used as a storage reservoir, dams being necessary at both ends as the slope of the channel floor is very gentle.

Locality 6. Julian Park (NZ 814010). The channel lost its identity here as the waters emptied into Wheeldale Lake. The sediments of its flat floor, built up to approximately 156 m O.D., completely choked the pre-existing river valley so that post-glacial drainage followed the new line cutting the rock-gorge at New Wath Scar (NZ 820006) and Hollins Wood. Here, massive sandstones and siltstones of the Saltwick Formation, with frequent plant debris, as well as the Eller Beck Formation, are exposed in the immature rock gorge. The gorge can be reached most easily at Mallyan Spout (NZ 825009), a local beauty spot signposted from the Goathland road.

ITINERARY VI

Locality 7. Goathland. Cross West Beck and continue to Goatland, noting the position of Moss Slack (intake at NZ 821001), which carried the overflow waters along the northern edge of Two Howes Ridge from Lake Wheeldale to Newton Dale. A peat-filled depression behind Goathland Church (NZ 832011) marks the final drainage stage along this feature. If time allows, proceed along the Pickering road for 1.5 km for a view of the intake of Newtondale, the direct glacial drainage channel which carried the meltwaters south across the unglaciated central tract to join Lake Pickering.

Locality 8. Eller Beck (Eller Beck Formation). Proceed northwards from Goathland through Darnholm, and ford Eller Beck at NZ 835022, a small car park being available on the north bank of the river. Proceed downstream, crossing the river at a suitable point to examine the type section of the Eller Beck Formation. The following section is adapted from Fox-Strangways (1892) using the terminology of Knox (1973):

	m
Eller Beck Formation	
Sandstone unit:	
Sandstone, hard at the top, flaggy at the base, with a few impressions of *Modiolus* and *Meleagrinella*. Strong ripple marks are present, with many trace fossils.	3 to 4
Shale unit:	
Ferruginous, silty shale.	1.2
Tough, sideritic mudstone, crowded with fossils.	0.12
Ferruginous shale.	1.5
Goathland ironstone unit:	
Sideritic mudstone, full of comminuted shells.	0.35
Saltwick Formation:	
Shales.	seen to 0.6

The list of bivalves found in the ironstone beds is long:– *Pholadomya, Pleuromya, Astarte, Cardinia, Nucula, Tancredia, Trigonia, Ostrea, Gervillia* and *Pinna*. The gastropods *Littorina* and *Turritella* also occur. However, much of the fauna is stunted, and the small bivalves and gastropods are preserved in dickite (a clay mineral), and poorly displayed. Downstream, massive sandstones dominate the Saltwick Formation at this locality, forming a substantial waterfall (Thomason Foss). The hillside north of Goathland is scarred by quarrying both of this sandstone and of the Cleveland Dyke, the latter no longer being exposed.

ITINERARY VII

South Bay, Scarborough, Cayton Bay and Gristhorpe Bay

J. K. Wright

O.S. 1:25,000 Sheet TA 08/09/18 Scarborough or
1:50,000 Sheet 101 Scarborough
G.S. 1:63,360 Sheet 54 Scarborough

This itinerary demonstrates Middle and Upper Jurassic sequences south of Scarborough. The route follows the beach and rock platform from Scarborough through Cornelian and Cayton Bays to Gristhorpe Bay (6 km). The going is fairly straightforward. Hard hats are strongly advised. **The excursion is best done on a falling tide.** Difficulties may be experienced in several places when the tide is in, though there is no danger of being cut off. In particular, it is rarely possible to gain access to South Bay, Scarborough, from the end of the promenade within three hours of high tide. High tide reaches up to the sea wall in the centre of Cayton Bay, though a detour over the cliff-top path is readily available. The itinerary is designed to be carried out by leaving Scarborough on foot and returning from Cayton by bus. A regular bus service is available. Alternatively, one may proceed from Scarborough by car or minibus, parking first at Cornelian Drive car park and using the cliff-top path to get to Locality 1A (Figure 21), returning to the vehicle from Locality 4, and then driving to Cayton Bay car park and rejoining the itinerary at Location 5A.

Stratigraphically the itinerary is less straightforward that that to the north of Scarborough. Several faults belonging to the Peak – Red Cliff Fault Zone run obliquely into the coastal area from Castle Hill southwards. As one crosses and recrosses the faults one jumps up or down the succession several times, making reference to the tables of strata (p.17, 19) necessary. The advantage is that one is able to see a substantial part of the N.E. Yorkshire Middle and Upper Jurassic successions within a short distance. Boulder Clay regularly slips down over parts of the lower cliff and this means that some sections described here may not be accessible during a particular visit.

Locality 1. South Bay, Scarborough (Scarborough and Scalby Formations).
1A. The south end of the promenade. From here descend to the shore. The rock platform consists of the tough, sideritic White Nab Ironstone Member of the Scarborough Formation. Bioturbation is marked, with networks of *Thalassinoides* burrows infilled with shell debris. There is extensive development of sideritic concretions at several horizons. Bivalves are abundant, particularly *Gervillella, Meleagrinella* and *Pleuromya*.

ITINERARY VII

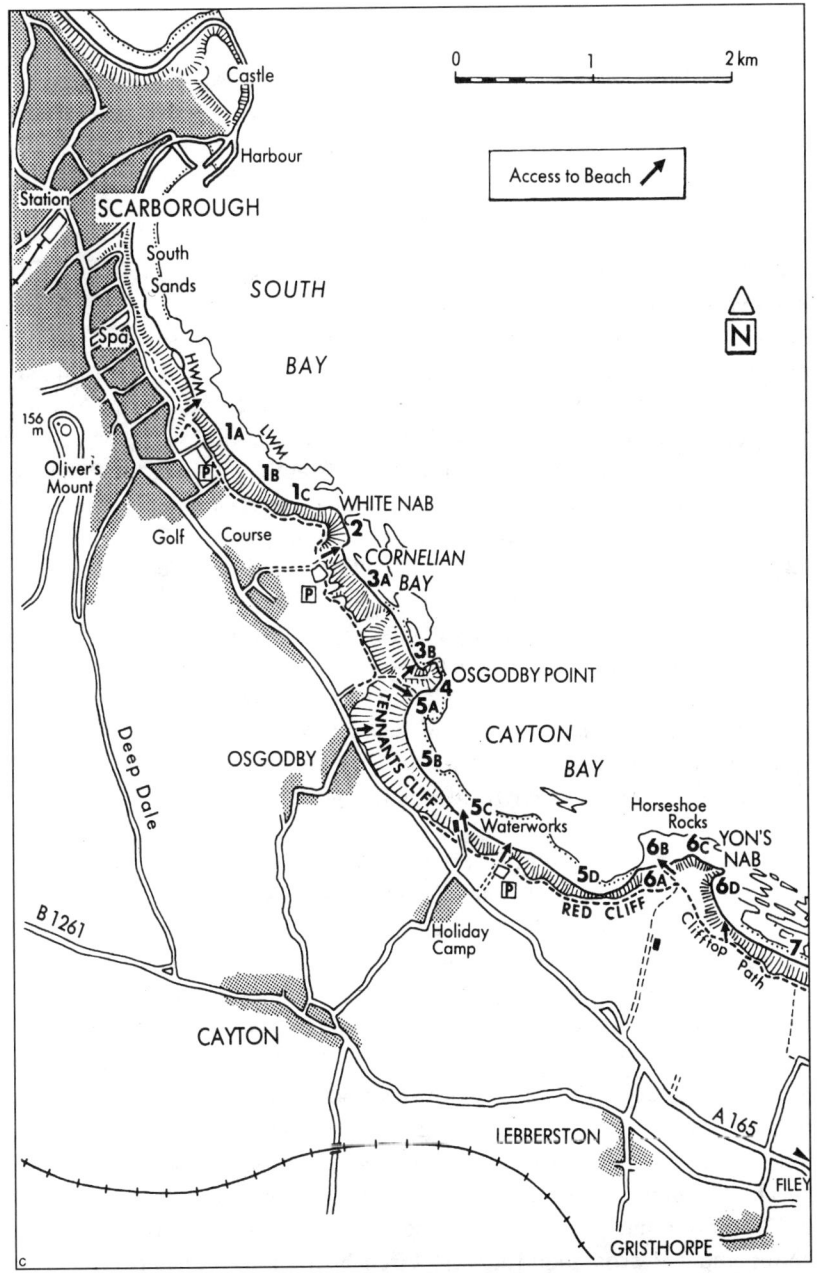

Figure 21. Map of localities in the Scarborough – Cayton Bay area.

Just above the base of the cliff, the grey, argillaceous Scarborough Formation is succeeded by the massive, laminated, cross-bedded sandstone of the Moor Grit. There is a 6 m sequence here of northwesterly dipping cross-bedded units produced by the gradual migration northwestwards of a substantial channel. Frequent chutes cutting through these point bar deposits have left small tough cross-stratified infills with bedding dipping largely in the opposite direction to the main cross-bedding.

1B. The centre of South Bay. Two hundred metres south of the promenade, the Scarborough Formation is overlain by cross-bedded channel sandstones dipping southwards (Figure 22). With favourable beach conditions, scouring of the uppermost Scarborough Formation can be demonstrated, the basal part of the overlying Moor Grit Member being coarse and ill-sorted. The persistent depositional dip in the lower cliff can be followed for several hundred metres southwards. Frequent chutes are filled with tough cross-stratified sandstone. The higher Moor Grit demonstrates deposition in a much less stable environment. Many channels are seen in section in the cliff (Figure 22). Most of the channels were infilled with sand and aggraded rapidly with little migration of the channel. Some channels were infilled with laminated silt and clay, and others became stagnant oxbow lakes infilled with plant debris and clay. The upper Moor Grit Member thus formed in an unstable environment. Periodic floods eroded the underlying beds and the stream channels aggraded rapidly as they became choked with sediment. Crevasse-splaying produced sheet sands alternating with alluvial silts and clays, and the abandoned channels were filled with fine grained sediment.

1C. The south end of South Bay. The southward migration of the channel laying down the persistent, southerly dipping, cross-bedded sandstone unit can be followed towards the southern end of the bay, where these sands pass laterally into the abandoned channel facies, with alternations of laminated, argillaceous siltstone and sandstone (Figure 22). South of this, erosion beneath the Scalby Formation has proceeded to a higher base level than to the north, and the interval between the White Nab Ironstone and Moor Grit Members is occupied by quasi-marine beds not seen at the previous localities in this itinerary. These form the Bogmire Gill Member of the Scarborough Formation (Gowland & Riding, 1991). Best seen at the far end of the sandy beach (2, Figure 22), the grey, shelly limestone described above is overlain by 2.1 m of white, level-bedded, fine grained sandstone showing delicate scour-and-fill cross bedding. At three horizons bioturbation is well developed. Some of the large pedestals on the upper rock platform show

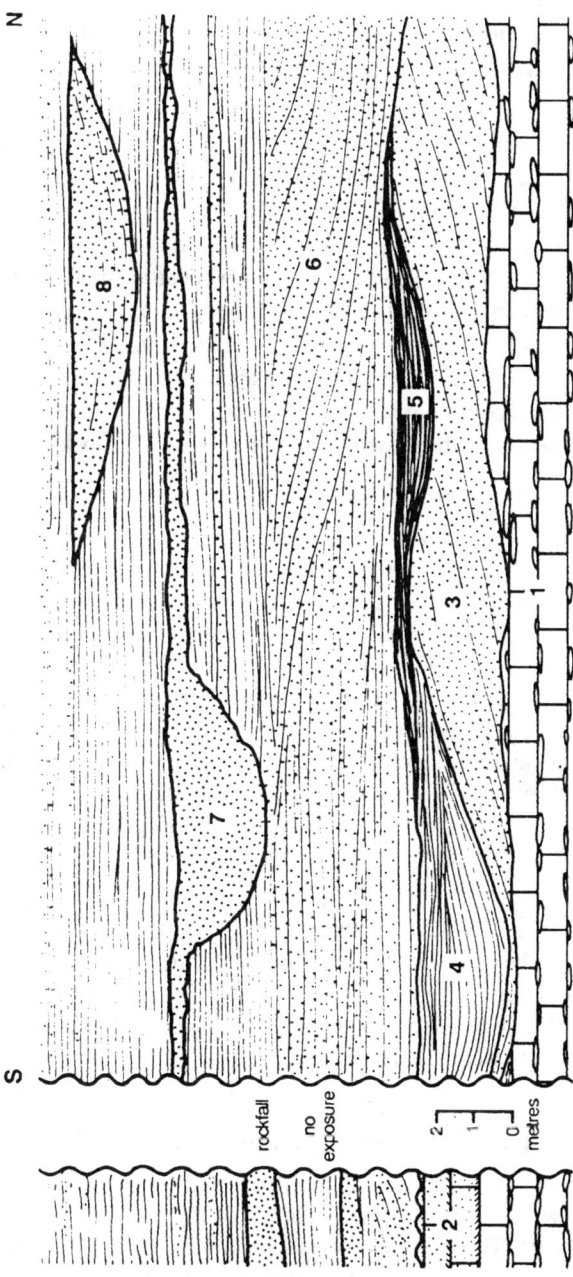

Figure 22. Schematic diagram to illustrate the main features of the Moor Grit and Long Nab Members as seen in the cliffs in the centre of South Bay. (1) White Nab Member. (2) Bogmire Gill Member (very fine grained marine sandstone). (3) Epsilon cross-bedded sands laid down by a mature river channel migrating steadily southwards. (4) Channel infill of silt, clay and fine sand. (5) Abandoned channel infilled with carbonaceous debris. (6) Rapidly aggrading channel migrating northwards. (7) Crevasse-splay channel rapidly eroded and infilled with unbedded sand. (8) Short-lived channel crossing the alluvial flats. Not all these features are seen in any one section.

beautifully the erosion of this fine grained sandstone, with hollows cut in its top surface infilled with the coarse sand of the Moor Grit. The Bogmire Gill Member was clearly well indurated before being eroded by the Moor Grit streams.

Eighty metres southwards, in the basal Moor Grit, is a bed of sideritic mudstone laid down in a small, abandoned channel, and from here the remains of the fish *Heterolepidotus* and turtle plates have been found (Woodend Museum Collection, Scarborough). The sloping upper cliff is formed of the Long Nab Member, 45 m thick, which consists chiefly of non-marine shale with occasional channel sands. These beds occupy the cliff up to the Cornbrash Limestone Formation exposed just beneath the Golf Course. Many blocks of Cornbrash Limestone containing abundant *Trigonia elongata* and *Myophorella scarburgensis* are found on the upper beach in the centre of South Bay. *Macrocephalites* can be found occasionally in these blocks.

Locality 2. White Nab (Scarborough and Scalby Formations).

At White Nab, a gentle anticlinal structure has brought up the Scarborough Formation sufficiently to enable a 5 m succession to be examined working out towards low water mark. The White Nab Ironstone Member comprises a 2.5 m succession of alternations of sideritic mudstone and calcareous, fossiliferous shale (Parsons, 1977). The best preserved fossils, including ammonites, occur in concretions in the shale horizons. A massive, 2 m thick sandy limestone was quarried northwest of here in Victorian times as building stone for the pier at Scarborough. Attribution of the lowest beds seen at White Nab is still uncertain due to the presence of small faults. Fossiliferous shales seen at low water mark beneath a thin, red-weathering limestone have yielded a variety of well preserved fossils including ammonites preserved within mudstone nodules.

In the base of the cliff at White Nab, the White Nab Ironstone is succeeded after a sharp break by shelly, bioturbated marine sandstones. This is followed by 3 m of gently cross-bedded fluvial sandstones, and then by strongly cross-bedded sandstones. The White Nab succession thus shows a steady, step by step progression from marine carbonates through to strongly cross-bedded fluvial sandstones.

Locality 3. Cornelian Bay (Scalby and Osgodby Formations).

3A. The centre of Cornelian Bay. Along most of Cornelian Bay the beds are almost horizontal, with the Long Nab Member of the Scalby Formation exposed in the low cliff. The member consists predominantly of shale when first seen, with thin sheets of sandstone developed at several levels, and occasional cross-sections through sand-filled channels. Progressing southwards towards the prominent channel sandstone which cuts across the beach in the centre of the bay, the relations between channels, siltstones and clays can be demonstrated in the cliff section. The clays pass laterally into alternations of clay with thin siltstones laid down during periodic flooding by the adjacent channel. These beds then pass into cross-bedded channel sandstones dipping gently north. A planar-bedded, crevasse-splay sandstone cuts across the earlier beds, which are highly contorted at the junction due to water escape. The prominent channel with its war time "Pill Box" then cuts right down onto the Moor Grit Member out on the rock platform. This massive, cross-bedded sandstone can itself be traced southwards into alternations of thick, sandy beds with laminated siltstones and sandstones.

3B. The southern end of Cornelian Bay. Some 200 m from the southern end of Cornelian Bay the western branch of the Cayton Bay Fault is crossed. The downthrow of 45 m to the east brings the Cornbrash Limestone Member down to beach level. The succeeding divisions of the Osgodby Formation, the Red Cliff Rock, Langdale and Hackness Rock Members, with the Oxford Clay Formation, are normally seen in the southeastern corner of the bay, though slipped boulder clay can be a problem. The upper ironstone beds of the Red Cliff Rock and the coarse chamosite oolite of the Hackness Rock (0.7 m) just beneath the Oxford Clay Formation yield occasional ammonites (Wright, 1968).

Locality 4. Osgodby Point (Millepore Bed, Lebberston Member).

The eastern branch of the Cayton Bay Fault, with an upthrow of 110 m to the east, cuts through the headland of Osgodby Point and brings the Millepore Bed up to form a natural barrier protecting the headland. In fact, the Eller Beck Formation is brought up along the line of the fault in the centre of Cornelian Bay, but this exposure is only visible at very low tide (Livera & Leeder, 1981). Proceed eastwards over the huge fallen blocks until the Millepore Bed is reached *in situ* at the point. (Alternative easier access to the point is over the headland into Cayton Bay and return along the rock platform beneath the cliff). The Millepore Bed can be examined in great detail along the numerous intersecting joint surfaces and on weathered blocks. It comprises a cross-bedded, bioclastic limestone, there being at least five courses of limestone, the cross-bedding in each having a preferential dip varying in direction from one course to another. The two highest beds have in their top surfaces extensive infilled *Thalassinoides* burrow networks suggestive of pauses in sedimentation, but the lower courses of limestone exhibit no such evidence. The Millepore Bed was thus laid down in very shallow water, with strong tidal currents sweeping in the coarse shell sand and producing the marked cross-bedding. A prominent constituent of the shell sand is the bryozoan *Haploecia [Millepora] straminea*, and this exposure should be regarded as the type locality for the Millepore Bed. In the cliff above, the Yons Nab Beds, which are at the base of the Gristhorpe Member, are represented by a barely marine sequence of flaggy, laminated sandstones and siltstones showing intertidal flaser structure and occasional bioturbation. They are succeeded by strongly cross-bedded channel sandstones.

Locality 5. Cayton Bay (Ravenscar to Corallian Groups).

5A. The northern end of Cayton Bay. Rounding Osgodby Point, the main Cayton Bay Fault is again crossed, and a small, boulder-strewn area intervenes before the sandy bay is reached. This area of rocks contains an interesting section through the Callovian strata. The beds dip steeply northwest into the low cliff, and are separated from the cliff by a small strike fault. The Cornbrash Limestone Formation forms the reefs at mid-tide level. It is fossiliferous, with bivalves and *Macrocephalites kamptus*. Cut laquered blocks show excellent bioturbation structures. The Cayton Clay Formation is sometimes covered by sand, but can yield phosphatic nodules containing the shrimp *Meyeria*, and also *Macrocephalites kamptus*, with authigenic zinc blende. In the Red Cliff Rock Member the iron-rich beds near the top have yielded well preserved keppleritid ammonites. The coarse, chamosite oolite of the Hackness Rock Member yields occasional interesting perisphinctids.

5B. Tenant's Cliff. Crossing the sands southwards now, the next stop is at the crumbling Lower Calcareous Grit cliffs of Tenant's Cliff. The Tenant's Cliff Member of this formation comprises a thick bedded, fossiliferous calcareous sandstone. Almost all the fauna has been collected from the numerous concretions exposed both in the rock platform to the north and east of the cliff and, across a small fault, in the cliff itself. Since the site was discovered early last century collectors have broken open the concretions to obtain the excellently preserved ammonite fauna of cardioceratids, perisphinctids and oppeliids with additionally bivalves, brachiopods and gastropods. It is rarely worth attempting to break open a concretion unless cross sections of fossils are visible on the outside; 90 per cent of concretions are barren.

5C. Cayton Bay Waterworks. At the southeastern end of Tenant's Cliff at low tide a section in the higher Middle and lower Upper Jurassic rocks is seen dipping gently landwards (for map see Wright, 1968, fig. 9). The Scarborough Formation is exposed at very low tide on the eastern side of the Cayton Bay Fault. When beach conditions are good there can be seen 5 m of fine grained, argillaceous limestone and calcareous sandstone containing occasional bivalves, overlain by 1 m of massive, medium grained shelly limestone. These clearly represent the Hundale Sandstone and Spindle Thorne Limestone Members. Their presence here is very significant, for they are not present only 1.5 km away at Yons Nab, almost certainly due to contemporary movement on the Red Cliff Fault (see Locality 6 below). Moving towards the Waterworks, the substantial fault with a 70 m throw is crossed, and the next beds seen are the Red Cliff Rock Member, yielding very occasional ammonites, and a thin remnant of the Langdale Member, yielding only bivalves. In contrast, the Hackness Rock Member is well exposed and yields numerous *Quenstedtoceras* spp. and *Peltoceras* spp. The Oxford Clay Formation is sometimes hidden by beach sand, but pyritised *Cardioceras* and *Peltoceras* can usually be collected.

5D. Red Cliff. Red Cliff is an imposing sight as it is approached across the wide sandy beach (Figures 23 & 24). The Osgodby Formation sandstones form the lowest quarter of the cliff, the steep slopes above are formed of the Oxford Clay Formation, while the Lower Calcareous Grit Formation forms the vertical face in the upper part. The very fossiliferous Cornbrash Limestone Formation forms a reef across the shore beneath Red Cliff, where a thin limestone with numerous *Lopha marshii* overlies 37 cm of chamosite oolite limestone. *Rhizocorallium* burrows from the base of the latter penetrate the underlying Scalby Formation siltstones (Wright, 1977). Unfortunately these beds are often hidden beneath beach sand or seaweed. Both the Cornbrash and overlying Cayton Clay Formation are sometimes exposed at the NW end of the cliff (TA 074842).

At the foot of Red Cliff the Osgodby Formation is accessible in places, **though falling shale from above makes it dangerous to work there**. The bulk of the formation here is represented by the Red Cliff Rock Member, for which Red Cliff is the type section (Page, 1989). Many fallen blocks of fine-grained chamosite oolite limestone from the upper Red Cliff Member occur on the upper beach, and yield abundant bivalves and occasional ammonites, especially *Kepplerites*. The Langdale Member is absent here due to intraformational erosion (Wright, 1968). The Hackness Rock Member comprises the 1-2 m of chamosite oolite limestone at the top of the Osgodby Formation, immediately beneath the Oxford Clay Formation. It has yielded occasional *Quenstedtoceras* and *Kosmoceras*. The Oxford Clay and Lower Calcareous Grit Formations (the latter often riddled with *Thalassinoides* burrows) are almost inaccessible, but fallen blocks show that they are largely barren, the latter surprisingly so considering that Tenant's Cliff with its prolific fauna is only 1 km away.

Locality 6. Yons Nab (Ravenscar Group and Red Cliff Fault).

Proceed now along the foot of the cliff, and over the undercliff by the path leading past a notice warning of the impossibility of walking all the way to Filey along the beach.

6A. Eastern end of Red Cliff. On reaching a small gully at the eastern end of Red Cliff, note that the Red Cliff fault runs down the gully, and that the Red Cliff Member is now at the top of the cliff on the eastern (upthrown) side of the fault (Figure 24). The throw here is about 37 m.

Yons Nab is composed of a gently, westerly dipping succession of Middle Jurassic strata lying beneath the Red Cliff Member. The outermost reef is formed of the Millepore Bed, and it and the Yons Nab Beds run straight across the rock platform to meet the fault as it runs out northwards into the sea.

All the overlying beds run into the cliff and rise gently as one proceeds east along the nab, and then descend back into the beach as one heads west into Gristhorpe Bay. The succession is thus described here as it is seen, starting at the top.

6B. Northern side of Yons Nab. The first beds one comes to in the low cliff section are Moor Grit Member sandstones and siltstones. There are irregular alternations of sandstones containing much carbonised wood with laminated, silty shales. The Moor Grit again shows evidence of fast sedimentation, with small-scale channelling followed by rapid infilling.

Below come the 3.3 m of the Scarborough Formation. An upper unit consists of 2 m of soft, argillaceous, shelly limestones with a layer of concretionary ironstone (White Nab Ironstone) near the top. *Lopha* is abundant. Below come 1.3 m of the Helwath Beck Member (Gowland & Riding, 1991), consisting of delicately laminated siltstone with the bedding picked out by carbonaceous layers. Ripple-drift bedding and scour-and-fill structures are beautifully displayed.

The Scarborough Formation is very attenuated at Yons Nab compared with the sequence, probably 12 m thick in all, present at the centre of Cayton Bay. The Hundale Shale, Hundale Sandstone, Spindle Thorne Limestone and Ravenscar Shale Members, most of which are present in Cayton Bay, are absent at Yons Nab. The Cayton Bay exposure lay within the Peak Trough (Milsom & Rawson, 1989), a fault-bounded trough operating in Middle Jurassic times which allowed thicker sequences of strata to accumulate within the trough than on its flanks. Particularly during deposition of the Scarborough Formation, subsidence on the eastern flank of the trough, east of the Red Cliff Fault, was much less than that in the centre. In addition the Scalby Formation, Cornbrash Formation, Hackness Rock Member and Passage Beds Member all show some attenuation in thickness in the area east of the Red Cliff Fault.

Figure 23 *(Upper)*. High Red Cliff, Cayton Bay. The vertical face in the lower part of the cliff is formed by the Osgodby Formation, the paler coloured slopes above are the silty clays of the Oxford Clay Formation, while the top part of the cliff is formed by the Lower Calcareous Grit Formation. *(Lower)*. The Scarborough Formation on the north side of Gristhorpe Bay. The figure is sitting on the siltstones of the Helwath Beck Member. Above are about 2 m of soft, argillaceous shelly limestones. The hard band in the upper part is the White Nab Ironstone.

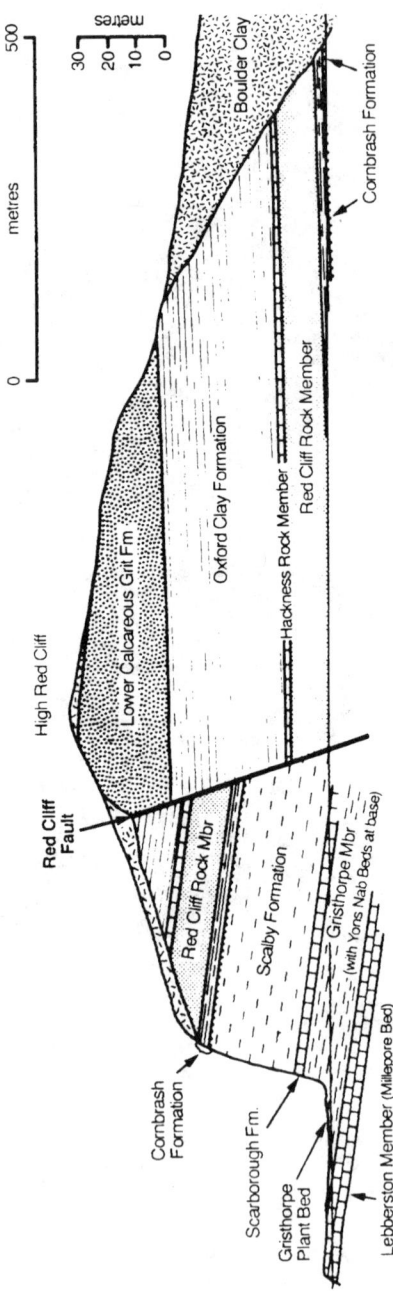

Figure 24. Cliff section at the south end of Cayton Bay.

ITINERARY VII 77

The silts of the Helwath Beck Member rest on 1.5 m of silty, carbonaceous shale resting on a rootlet siltstone (0.7 m). Beneath the rootlet bed is the well known Gristhorpe Plant Bed (2.5 m), consisting of thinly laminated, friable clays containing abundant plant debris and well preserved stems and leaves of Bennetitales, Ginkgoales, Conifers, Ferns, Pteridophytes and Caytoniales. The plant bed rests on some 4.5 m of largely marine sandstones. These are tough, with iron-rich concretions below, strong bioturbation about the middle, and with irregularly developed bioturbation towards the top. The lower 1.5 m of this sandstone, which is fossiliferous, was included within the Yons Nab Beds by Bate (1959).

6C. Horse Shoe Rocks. All the beds described above can be followed from the rock platform into the cliff. The Yons Nab Beds run across the rock platform just east of the nab. They consist of 5 m of flaggy alternations of shale and siltstone with numerous small, sideritic concretions and a marine bivalve fauna. The Millepore Bed forms the massive rampart at the seaward end of the rock platform. The highest 2 m consist of oolitic limestone with cross-bedding detectable beneath the barnacle-encrusted surface. Beneath, 7 m of cross-bedded, calcareous sandstone is seen at very low tides, resting on fluvial channel sandstone of the Sycarham Member.

6D. North end of Gristhorpe Bay. The whole sequence can now be followed back in stratigraphic order into Gristhorpe Bay. Of particular note are exposures of the Scarborough Formation in the cliff (Figure 23) and rock platform displaying an abundant bivalve fauna. A large channel in the Moor Grit Member strongly cross-cutting level-bedded siltstones, seen at the east end of the nab, was first illustrated by Black (1928, fig. 1).

Locality 7. Gristhorpe Bay (Cornbrash and Osgodby Formations).
Proceed now to the centre of Gristhorpe Bay. A sequence from the Scalby Formation siltstones to the Oxford Clay Formation can be examined in the cliff, but the chief interest of this exposure lies in the variety of fallen blocks from which a very varied fauna can be collected. The three thin limestone units of the Cornbrash Limestone Formation can be distinguished: the basal, brown-weathering, sideritic limestone, the middle pale grey fine grained micritic limestone containing *Trigonia* and many gastropods, and the upper bioclastic limestone containing bivalves and *Macrocephalites*. This is the best locality for Middle Callovian Langdale Member ammonites, the fallen blocks of fine to medium grained sandstone yielding *Erymnoceras, Perisphinctes* and *Kosmoceras* (Wright, 1968). The Hackness Rock Member is a pale grey limestone with scattered chamosite ooliths and yields *Kosmoceras spinosum, Quenstedtoceras* spp. and *Collotia* sp. Blocks of Oxford Clay yield well preserved if flattened *Cardioceras praecordatum* and *Parawedekindia arduennensis* (Wright, 1983).

ITINERARY VIII

Filey Brigg

J. K. Wright

O.S. 1:25,000 Sheet TA08/09/18 Scarborough or
1:50,000 Sheet 101 Scarborough
G.S. 1:63,360 Sheet 54 Scarborough

This is a half day excursion examining the lower part of the Coralline Oolite Formation (Passage Beds, Hambleton Oolite and Birdsall Calcareous Grit Members) well displayed in the wave-cut platform and low cliff of the Brigg. The going is fairly straightforward. Parking is available in the cliff-top car park (TA 119811).

The map (Figure 25) shows the geology of the area and the recommended stopping places. The log brings out the predominantly sandy nature of the succession. Without three noticeable beds of oolitic limestone within the Lower Leaf of the Hambleton Oolite Member it would have been very difficult to subdivide the succession or to correlate it with that seen over the rest of the Cleveland Basin. The correlation here follows Wright (1983) in referring the massive sandstones towards the top of the succession to the Birdsall Calcareous Grit Member rather than to the Middle Calcareous Grit Member of previous correlations. The Birdsall Member comprises a wedge of sand poured into the southern side of the Cleveland Basin during an uplift of the Market Weighton High. Oolite (Hambleton Oolite Member) continued to be deposited throughout in the northern half of the basin, but where the Birdsall Member is present the Hambleton Oolite is divided into Lower and Upper Leafs.

A. North side of the Brigg. Proceed along the cliff-top path to the end of the Brigg and descend to the shore. Continue over the rock platform and boulders to the small bay 100 m NW of the hut at the end of the path. The rock platform here displays the round "cannonball" concretions so typical of the Saintoft Member of the Lower Calcareous Grit Formation. Resting on an erosion surface cut in the Saintoft Member, the lowest 0.6 m of the Passage Beds Member consist of a fine to medium grained sandstone. The sandstone is heavily bioturbated and contains *Nanogyra* and *Chlamys*. Above comes the main Passage Bed limestone, 2 m of grey-weathering limestone in 5 or sometimes 6 beds (for faunal list see Wilson (1949), beds 1-8). *Nanogyra* colonies weather out, and there are many dissociated *Gervillia* valves in the upper beds, and some shells in life position in the top bed. Small-scale cross-bedding fills small scours, and dips to the south. The Passage Beds seem to have been deposited in a series of storm surges washing shell debris from shallow water to the northwest into the offshore shelf area to the southeast. Only in the highest bed is the fauna indigenous. Much of the time there appears to have been little or no sedimentation. In Itinerary IX, 12 m of Passage Beds will be examined near Pickering.

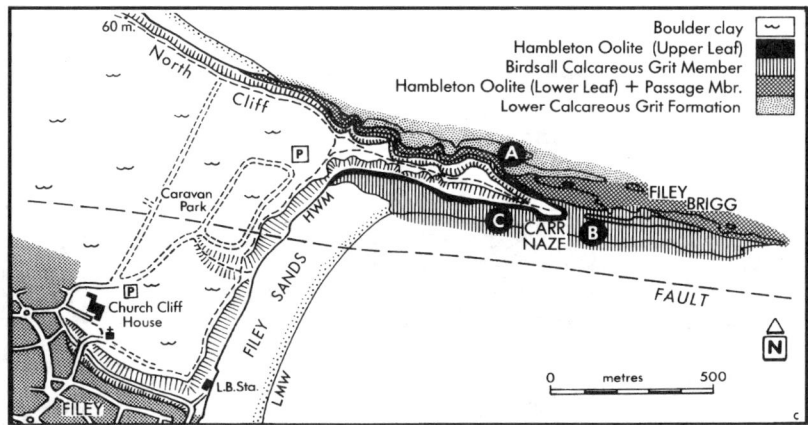

Figure 25. Sketch map of the geology of Filey Carr Naze and Brigg.

The next 3.7 m of limestone belongs to the Lower Leaf of the Hambleton Oolite Member. A major bedding plane marks the base, above which comes the first massive bed of oolite. None of the oolites at Filey show cross-bedding and they seem to have accumulated in a fairly stable, quiet environment allowing the excellent preservation of delicate echinoids and brachiopods (for faunal list see Wilson (1949) beds 9-14 only). The quiet conditions also favoured the development of extensive networks of *Thalassinoides* burrows. The burrow systems weather out in spectacular fashion in the large, fallen blocks towards the centre of the Brigg (Figure 27). The higher, sandy limestones of the Lower Leaf form the NW side of the Brigg.

B. Filey Brigg. The Brigg itself is formed of a tough calcareous sandstone, the Birdsall Calcareous Grit Member. The full thickness is 6.8 m. Towards the base there are calcareous concretions with shelly bands containing occasional *Cardioceras* spp. Massive, occasionally cross-bedded sandstone forms the bulk of the unit with, at the top, two beds of tough calcareous sandstone. Fossils from these beds can be collected loose from the gravel and boulders on the south side of the Brigg.

C. South side of Filey Brigg. The Upper Leaf of the Hambleton Oolite Member is excellently seen in a continuous section along the south side of the Brigg. It consists of tough, impure, very fossiliferous, bioclastic limestone containing well preserved bivalves and occasional *Cardioceras* and *Perisphinctes*. *Perisphinctes* is not found in the Lower Leaf and Birdsall Calcareous Grit below, these members belonging to the *cordatum* Subzone. The presence of *Perisphinctes* in the highest beds suggests the presence of the *vertebrale* subzone of the *plicatilis* Zone, normally indicative of the very highest Hambleton Oolite Member. Only 1.5 m is seen, however, beneath the limestone rubble at the base of the boulder clay.

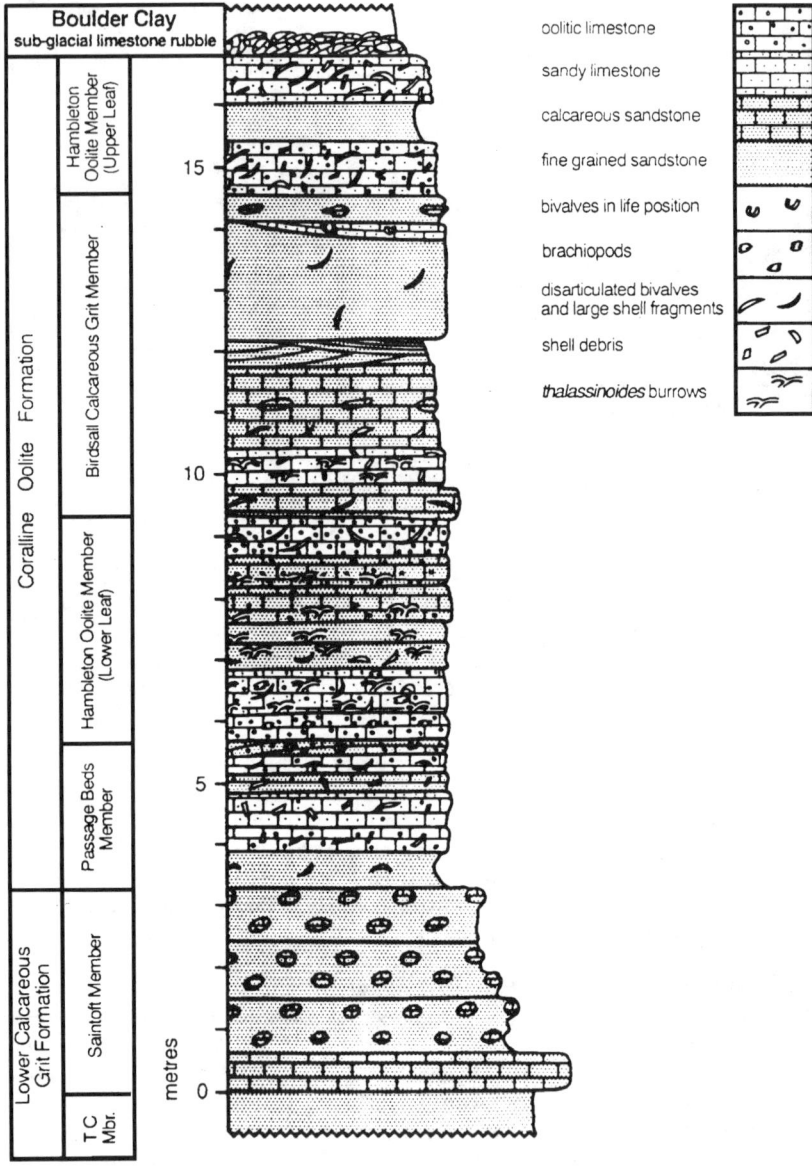

Figure 26. Section through the Corallian at Filey Brigg.

Figure 27 *(Upper)*. The Corallian sequence north of Filey Brigg. The ledges in the foreground and in the lower part of the cliff form the upper part of the Lower Calcareous Grit Formation. The prominent bedding plane at the foot of the ladder marks the base of the Coralline Oolite Formation. *(Lower)*. *Thalassinoides* (crustacean) burrow systems etched out on a loose block from the Lower Leaf of the Hambleton Oolite Member.

ITINERARY IX

Castle Hill, Scarborough, and the Hackness Hills

J. K. Wright

O.S. 1:25,000 Outdoor Leisure Map Sheet 27, and TA08/09/18 Scarborough
1:50,000 Sheets 100 Malton & Pickering and 100 Scarborough
G.S. 1:63,360 Sheets 35 with 44 Whitby & Scarborough and 54 Scarborough

The itinerary (Figure 28) is designed to demonstrate the relationships of sedimentary facies and the fossils contained in the varied series of limestones and sandstones which make up the major Coralline Oolite subdivision of the Corallian Group in NE Yorkshire. The itinerary is arranged as a tour through the Hackness Hills starting and finishing at Scarborough. The going is easy. Private transport, either car or minibus, is essential. Coaches are unsuitable.

Locality 1. Castle Hill, Scarborough (Corallian Group).

Free parking may be available at the top of Castle Road just east of St Mary's Church, though this is unusual during the summer season. Alternatively, one can pay to park on the Marine Drive in the North Bay and walk up to the castle. From St Mary's Church proceed along the footpath to the right of the road leading up into Scarborough Castle and take the second path on the left through the archway leading to the North Side. If starting from the Marine Drive proceed up paths to the archway which is clearly visible. Then proceed along the rough, unmade path leading eastwards below the Castle walls.

1A. The Holms. After 100 m, a beautifully weathered rock face is reached. This shows the Saintoft Member of the Lower Calcareous Grit Formation capped by Passage Beds sandstone. There are very fine arboresques consisting of ramifying networks of infilled *Thalassinoides* burrows and strongly weathered calcareous concretions. The popular name "ball beds" is readily evident. Fallen concretions have yielded fragmentary bivalves and ammonites (Wright, 1983). A tough, calcareous sandstone marks the base of the Saintoft Member.

1B. Castle Cliff. A further 100 m to the east, a more complete section is reached, extending from the Lower Calcareous Grit Formation through the Passage Beds into the base of the Hambleton Oolite Member. A measured section was given by Wright (1983). Below the path, the Tenant's Cliff Member of the Lower Calcareous Grit, though well exposed, is poorly fossiliferous. The Saintoft Member is largely hidden by grass. However, the Passage Beds sequence is well displayed and can be examined in detail. After an initial 1.5 m of fine to medium grained, shelly sandstone the Passage Beds then comprise 7 m of alternations of laminated, silty clay with cross-bedded,

ITINERARY IX   83

shelly limestone. The juxtaposition of two such contrasting facies may be explained as follows. The silty clays appear to have been laid down in deeper water to the east of the shallow shelf sea occupied by the Hackness coral-sponge reef (Locality 2). The cross-bedded shell sands were swept into the Castle Hill area during storms. The dominant cross-bedding direction is thus eastwards, with subordinate westerly dips. Further up the succession the clay bands become thinner and encrusting mats of *Nanogyra nana* appear, showing that at least part of the shell fauna was indigenous rather than transported. The highest shelly grainstones (limestones) of the Passage Beds, containing 10 per cent comminuted coral debris derived from the reef, show markedly undulating bedding, and it is difficult without the aid of thin-sections to decide where the change to fine oolite at the base of the Hambleton Oolite lies. Much more typical of the latter is the massive-bedded oolite exposed immediately below the Castle Walls, with numerous reworked *Gervillia* valves weathering out.

Locality 2. Hackness Head (Passage Beds Member).

Proceed west now to the village of Hackness, 8 km from Scarborough (Figure 28). Hackness Head Quarry (SE 966904) is on private land, and permission to visit must be obtained beforehand via letter (enclosing SAE) or 'phone call from Mr C. Swiers, Broxa Farm, Hackness, Scarborough, YO13 0BP, tel. 0723 82273. Park at SE 967903 in the straight stretch of road southwest of Hackness Church. Proceed up the slanting path leading westwards up the hillside, take the right fork, and immediately turn right up a steep track. Continue up by the rough, overgrown path leading into the small field at the east end of Hackness Head.

There are two small quarries here. The eastern one shows 3.7 m of the shelly, iron-rich facies of the middle Passage Beds. *Nanogyra nana* is abundant, with occasional *Chlamys* and sponges. Some coralliferous limestone occurs at the top of this section, but it is best exposed in the western quarry, which reveals two metres of white, rubbly-weathering coral-sponge limestone, the earliest known reef from the British Corallian. In the talus below the low face of the quarry can be found large colonies of *Thamnasteria arachnoides* and fragmentary *Thecosmilia annularis*. In the finer gravel are the delicate branching coral *Rhabdophyllia phillipsi* with many *Holcospongia* sp. figured by Wilson (1949, pl. 11), and also terebratulids. The reef thus consisted of discrete colonies of *Thamnasteria*, with the spaces in between occupied by sponges, brachiopods, bivalves and echinoids. Shelly lime mud was deposited between the coral colonies, much being derived from the activities of coral-boring organisms such as *Lithophaga*. New coral colonies grew on the bored remains of the old. In sheltered areas the more delicate stags horn corals *Thecosmilia* and *Rhabdophyllia* grew profusely. This was clearly a back-reef, lagoonal environment. The higher energy reef-top and reef-apron environments are not seen, the Hackness reef as it exists today being only a remnant of the former extent of reef facies.

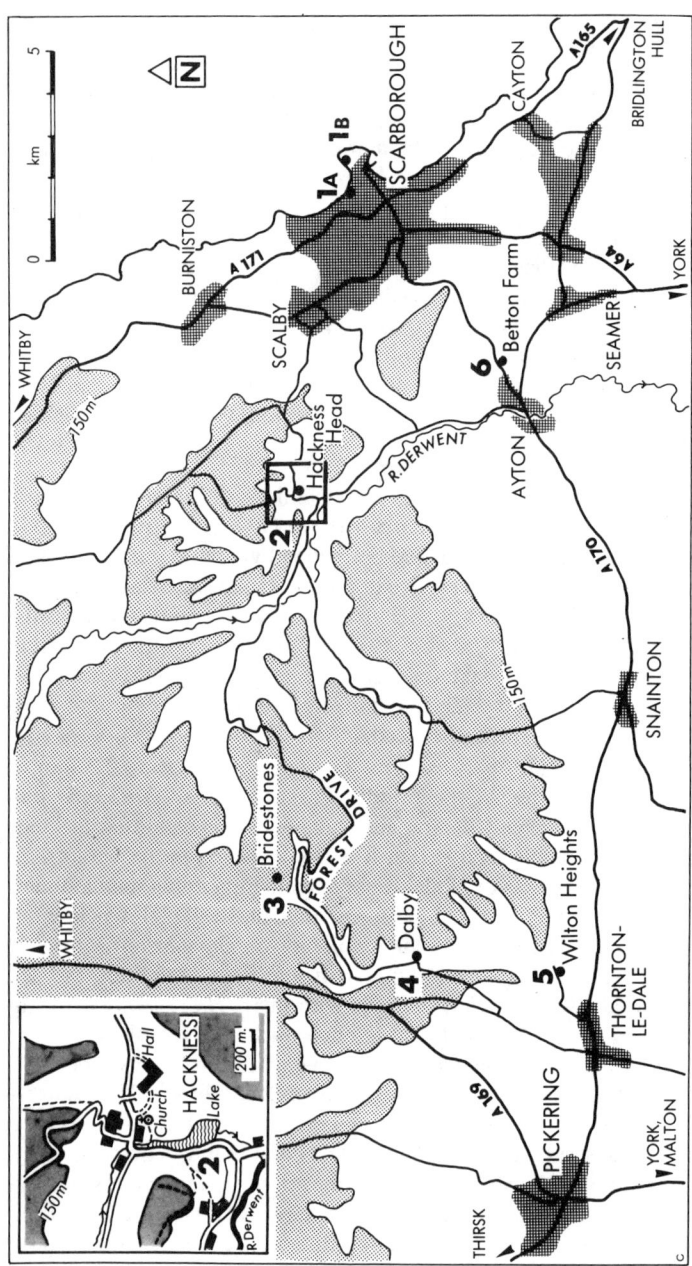

Figure 28. Map of localities in Scarborough and the Hackness Hills.

Locality 3. The Bridestones (Passage Beds Member)

This visit to the naturally weathered tors in Passage Beds sandstone occupies approximately 1 hour. From Hackness, follow the Langdale End road passing through this village and on to Bickley. Here the Forest Drive begins (a toll is currently charged). From the car park in Staindale (SE 878906) proceed NNW along the track marked Nature Walk, and along the path which slants NNW up through the trees.

The origin of the Bridestones has given rise to a certain amount of controversy in geographical circles. Palmer (1956) considered that during a phase of rejuvenation of the valley slopes in the last interglacial, a natural scarp formed running along the upper valley sides where the Passage Beds outcropped. This scarp was modified subsequently by prolonged erosion and weathering along joints to leave the pedestals or tors of calcareous sandstone remaining, marking the line where the scarp originally ran.

Linton (1955) in what is a more likely explanation held that the Bridestones could not have formed in this one stage weathering process, and that a two stage process was necessary. In his view deep weathering along joints during the last interglacial left broad zones of loose, weathered rock and sand at and close to the surface. In between the weathered zones were residual, unjointed masses of unweathered bedrock. Mechanical stripping off of the weathered materials during the solifluction stages of the last glaciation left pillars of unweathered rock standing along the valley sides where weathering has been deepest and erosion most active.

The facies of the Passage Beds at the Bridestones is very unlike those seen at the last locality. It comprises a hard, calcareous, sparsely shelly sandstone showing marked cross-bedding with dips predominantly to the west or southwest. That this sandstone formed contemporaneously with the Hackness coral-sponge reef is shown by the frequent occurrence of coral fragments in sectioned blocks of it collected from Locality 4. The quartz sand may have been brought into the area by longitudinal transport along a beach barrier extending from the Market Weighton area to the south northwards into the deeper water of the central Cleveland Basin.

Locality 4. Dalby Cutting (Passage Beds).

The next locality, a roadside section (SE 854863), shows similar features, but there is no prohibition on collecting samples as there obviously was at the Bridestones. Proceed south down Staindale, through Dalby, and southwest up the opposite side of the valley. Near the crest of the hill, a continuous section is seen extending from the Saintoft Member of the Lower Calcareous Grit Formation through the Passage Beds into the basal Hambleton Oolite Member.

The Passage Beds Member is largely arenaceous but with increasing shell content upwards. Thus the lowest 3.9 m comprise calcareous sandstone, the middle 3.7 m shelly sandstone, and the upper 4.4 m are best described as a shelly, sandy limestone. The upper unit is strongly cross-bedded, the cross-sets in the main facing south, with minor north-facing cross-sets. A very shallow water near-beach environment is indicated. Coral fragments are common in this unit. *Aspidoceras* sp. has been collected here from the highest Passage Beds. The Hambleton Oolite rests on an erosion surface cut in the Passage Beds. Half a metre of oolite full of *Gervillia aviculoides* and *Chlamys fibrosus* is seen at the top of the section.

Locality 5. Wilton Heights Quarry, Thornton Dale (Hambleton Oolite Member).

Continue along the Forest Drive and south to Thornton Dale. Turn left along the A170 and at the top of the hill on the outskirts of the village turn left and proceed along a narrow lane for 2 km to the quarry entrance (SE 860843). The quarry is owned by Tilcon Ltd., and permission to visit must be obtained beforehand using the following procedure: 1. A written request must be made to the Quarries Product Manager, Tilcon Ltd., P.O. Box 5, Fell Bank, Birtley, Chester-le-Street, Co. Durham DH3 2ST. This must be at least 21 days before the proposed visit. 2. The official 'Letter of Indemnity' which will be forwarded by Tilcon must be completed and returned to the Quarries Product Manager at least seven days before the proposed visit. 3. Hard hats and high visibility clothing must be worn at all times during the visit, which is made entirely at the visitors' risk.

Most of the inland exposures of the Coralline Oolite Formation are in large working quarries where access for parties is difficult. Unused quarries deteriorate rapidly. However, Wilton Heights Quarry still has much to offer, as it was worked until quite recently. The quarry exposes 10 m of white, creamy oolitic limestone in even-bedded units with very little sign of cross-bedding. There are abundant disarticulated bivalves, including *Gervillia aviculoides*, *Myophorella* sp., *Trigonia reticulata*, *Chlamys fibrosus*, *Liostrea* sp. and *Camptonectes lens*. The gastropods *Cylindrites* sp. and *Pseudomelania heddingtonensis* are also common. The occasional belemnite guard is the only indication of cephalopods. The molluscs are more common in discrete bands of well sorted oolite and appear to have accumulated by the sweeping of comparatively undamaged shell debris into deeper water during storms.

In the centre of the main north-south face a small reef-like structure (bioherm) is clearly seen in cross-section. Blocks which have fallen down reveal that the bioherm began as a colony of *Nanogyra nana* growing on a small patch of firm substrate. *Thamnasteria arachnoides* then colonised this patch, and layers of *Nanogyra* and *Thamnasteria* built up to form a solid mass of limestone up to 1 m thick. During subsequent compaction the bioherm has been pushed down into the softer, yielding oolite beneath and around it. The colonial corals are much recrystallised and are riddled with the borings of *Lithophaga*. Occasional more delicate *Rhabdophyllia* also occur preserved in coarse oolite, and show the relatively quiet conditions under which the oolite accumulated. Small coral colonies and *Nanogyra* nests can be found elsewhere in the quarry. Numerous loose blocks of Middle Calcareous Grit occur, but the member is not seen *in situ* here. **Parts of the quarry face are crumbling and dangerous and should not be approached.**

Locality 6. Betton Farm Quarry (Malton Oolite and "Coral Rag").

The quarry is situated beside the busy A170 between Ayton and Scarborough. It will be necessary to park on the verge on the N.W. side of the road (TA 002857). Livestock are normally kept in the field in which the quarry lies, and visitors *must* obtain permission from Betton Farm, 200 m down the road, before entering the field. The quarry face has been cleaned up recently by the Nature Conservancy Council, and the following section was visible in 1989:

m

2. Shelly, micritic limestone containing fragmentary corals, echinoid spines, bivalves and gastropods, with large isolated masses of *Thamnasteria* up to 1 m across.   seen to 1.5

1. Well to massive bedded, very poorly sorted oolite containing *Bourgetia striata*   seen to 2.2

Bed 2 is extremely variable in its nature. Around the outside of the coral masses is a dense-packed oolith-coral-shell sand with abundant abraded fragments of massive corals. In between the coral stacks is a shelly, coralliferous, micritic limestone with abundant delicate coral fragments, including *Rhabdophyllia*, plus delicate bivalves and abraded coral fragments. The succession at Betton Farm Quarry thus represents a true reef complex, with channels choked with coral-shell sand separating large patches of coral growth. Within these patches finer, low energy sediment was deposited and fragile corals, bivalves, gastropods and echinoids could exist.

It must be emphasised that the coralliferous bed here is not the true Coral Rag as surmised by most authors. Hudleston (1877) was the first to note that the characteristic Coral Rag echinoid *Cidaris florigemma* is not present here, and that bed 2 is simply a coralliferous facies of the Malton Oolite Member. This was confirmed during the recent cleaning up of the quarry face, when it became evident that poorly sorted oolitic limestone infilled borings and crevices in the top surfaces of the coral patches, and that there was a return to the standard Malton Oolite facies in the highest beds.

ITINERARY X

Reighton Gap to Speeton Cliffs

P. F. Rawson

O.S. 1:25,000 Sheet TA 07/17 Hunmanby or
1:50,000 Landranger 101 Scarborough
G.S. 1:63,360 Sheet 55 Flamborough

Just north of Reighton village (TA 128757) turn off the coast road (A165) onto a minor road signposted for Reighton Sands holiday village. Follow this road until it forks at the holiday camp, then take the left branch to a small parking area (Figure 29; TA 140763). From here take the path to the beach and turn right towards the chalk cliffs in the far distance. There is a second access to the shore down a private road from the holiday camp. The shore is generally sandy, but occasionally patches are stripped off to expose either boulder clay or disturbed Kimmeridge Clay. The adjacent cliffs are of brown and reddish coloured boulder clay and show numerous landslips.

Locality 1. Middle Cliff to Speeton Beck (Kimmeridge and Speeton Formations).

1A. Middle Cliff. About three quarters of a kilometre from the footpath are a cluster of concrete blocks and a breakwater (Figure 29, BW1), in the vicinity of which the topmost paper shales of the Kimmeridge Clay are sometimes visible at the cliff foot or on the adjacent shore. Rarely, a band of large septarian concretions with attractive greenish-yellow calcite crystals is exposed. Flattened ammonites and small bivalves (*Lucina minuscula*) occur in the shales, which represent the *hudlestoni* to lower *pectinatus* Zones (Table III).

From here dark grey clays appear from beneath the boulder clay in the cliff, marking the commencement of the outcrop of the Lower Cretaceous Speeton Clay Formation. This extends along Middle and Black Cliffs for about 0.75 kilometre southeastward to Speeton Beck. The cliffs are unstable and continuously changing, so that rarely is the whole extent cleanly washed by the sea; instead there are often slipped clays along the cliff foot which can be very treacherous in wet weather. Conversely, patches of shingle and sand are sometimes stripped off the intertidal zone to give very clear exposures of the clays beneath. In both cliff and shore the clay shows small, sometimes tight folds formed by slipping, and even some of the faults that cross the beach may represent major landslip surfaces.

The Speeton Formation was divided by Lamplugh (1889) into 4 units, the A-D beds (labelled from the top downward: Table IV), characterised by abundant belemnites of alternating northern (Boreal) and southern (Tethyan) origin:

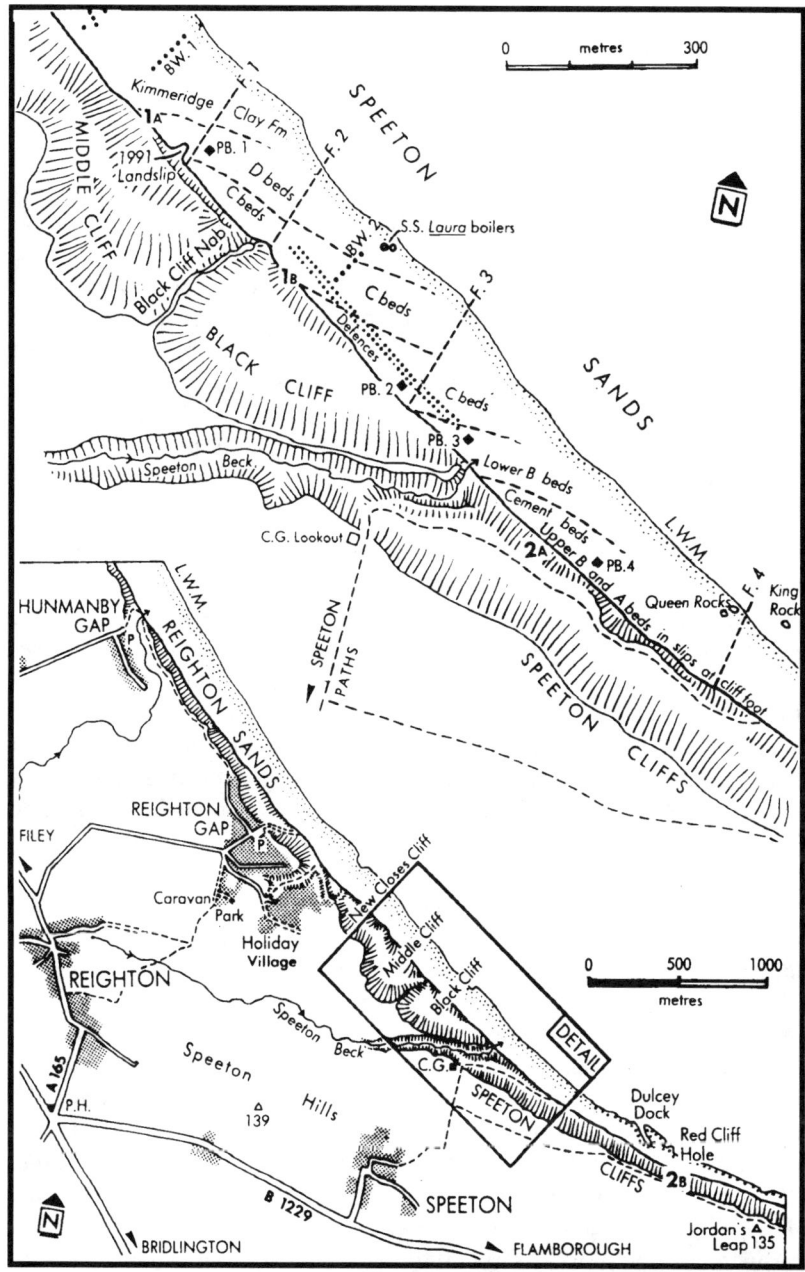

Figure 29. Map of the Speeton section, based mainly on the author's observations.

A beds – *Neohibolites* (T)
B beds – *Praeoxyteuthis, Aulacoteuthis* and *Oxyteuthis* (in ascending order) (B)
C Beds – *Hibolites* (T)
D Beds – *Acroteuthis* (B)

Finer lithological units were distinguished by Lamplugh and subsequent workers (see Rawson *et al.*, 1978 and Rawson & Mutterlose, 1983 for summaries). Although it is difficult to follow the section in detail unless it is very cleanly exposed, there are several distinctive marker horizons which one can look for as a key to reading the sequence. The first is at the base of the clays, where a 10 cm thick phosphate nodule bed (Coprolite Bed) with derived Kimmeridge Clay fossils marks reworking and a break in deposition over some 9 million years. This bed was mined as a source of phosphate until a major landslip closed the mines in 1869 (Lamplugh, 1889). Pit props are still uncovered occasionally at the cliff foot and help to locate the bed.

The commencement of deposition of the Speeton Formation is the local reflection of an important sea-level rise that flushed out the whole North Sea Basin (Rawson & Riley, 1982). From then on sedimentation continued slowly, with occasional interruption, through the remainder of the Lower Cretaceous.

If the Coprolite Bed is not visible the first obvious marker is usually the pale but bright, striped clay of D6. The overlying bed D5 is a brackish-water deposit which contains the primitive brachiopod *Lingula,* often preserved in its pyrite-infilled burrow. D4 marks a return to fully marine conditions and contains numerous bivalves, mainly *Astarte senecta* and the massive oyster *Exogyra latissima.* A brown-weathering silty clay with irregular nodules marks the top of D3 (D3A) and sometimes yields partially crushed large *Polyptychites.* A very distinctive band of large calcareous concretions enclosing smaller phosphatic nodules, the "Compound Nodule Bed" (D1), reaches the shore close to a large slab of concrete that once formed the base of a Second World War pillbox (Figure 29, PB1). Bed D1 is a condensed horizon rich in fossils, including *Acroteuthis*, ammonites (*Endemoceras* and *Distoloceras* up to 0.7 m diameter) and the "shrimp" *Meyeria ornata*. Only about a metre below D1 is a phosphatic nodule remanié horizon marking an important break in the sequence; the whole of the Upper Valanginian substage is cut out but represented by corroded ammonites among the nodules (Figure 30). Above D1, beds C11-C8 are quite fossiliferous, with attractively preserved small ammonites (including *Endemoceras regale, Olcostephanus* and *Parastieria peltoceroides*), *Hibolites jaculoides* and *Meyeria ornata*. A new ammonite fauna appears in C7, where *Simbirskites* (subgenus *Speetoniceras*) is common at the base and the uncoiling (heteromorph) ammonite *Aegocrioceras* (Figure 30) just above.

About 50 metres along the cliff from D1 is a thin, bright yellow sulphurous band about 3 m above the shore that forms a very clear marker zig-zagging along the remainder of Middle Cliff. This has weathered from a very thin pyritic layer within bed C7E and is not visible when the clay is freshly exposed on the shore. Just beneath it is bed C7F, the first of several silty, reddish-weathering bands that form distinctive markers higher up the succession. The first three (C7F, C7A and C5L) can be reached via gullies in the cliff. C7F contains common body chambers, sometimes with inner whorls, of *Aegocrioceras quadratum*, while C7A yields *Crioceratites*. The clays of C6 contain small *Simbirskites*, especially near the base and top.

1B. Black Cliff. Middle Cliff terminates at a ridge closely coinciding with a fault (F2) that crosses the shore just before the start of a line of concrete blocks paralleling the cliffs. To the SE of the fault a low vertical cliff exposes the upper C and lowest B beds, a series of thinly bedded, cyclic units in which the 10 cm thick, intensely glauconitic bed C2D is a good marker, as are the bioturbated clays of C1. Fossils are not common but the section is interesting for the occurrence of the small brachiopod *'Terebratulina' martiniana* in the glauconitic clays of bed C2A (see Middlemiss, 1976, p. 75).

Along Black Cliff there have been major landslips over the last ten years and the section through the Lower B Beds and basal Cement Beds detailed by Rawson and Mutterlose (1983) is largely obscured. However, the junction between Lower B and the Cement Beds is visible nearly opposite a second breakwater of concrete blocks (BW2). Here, very dark, laminated, kerogen-rich shaly clays are visible just beneath a distinctive double cementstone band that marks the base of the Cement Beds. These clays can be traced across the North Sea to North Germany where they are called the "Blatterton" (= paper shale). They represent a brief "anoxic" event in early Barremian times.

Where breakwater 2 runs out to sea two ship's boilers form a distinctive landmark. They form part of the wreck of the 2089 ton Austrian steamship Laura, which was sailing from Newcastle to Trieste with a cargo of coke when she ran ashore in dense fog on 21 November 1897 and broke in two (Godfrey & Lassey, 1974, p.73, photos, pp. 114-115). The wreck is sometimes exhumed from the sand, when both halves of the ship can be seen at very low tides.

Beyond the basal Cement Beds there is an extensive slipped area of boulder clay and red chalks. The Speeton Clay reappears again 30 m SE of a tilted pillbox (PB2), where the highest C Beds (brought up by Fault F3) are overlain by a continuous section through the Lower B Beds and, at Speeton Beck, the lowest Cement Beds. Although the section is currently (1991) much obscured by slip the sea has begun to uncover some beds again, and

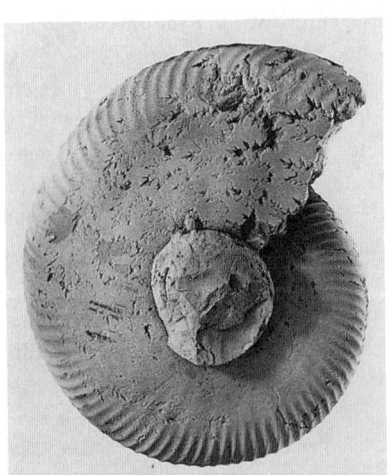 

Figure 30 *(Upper)*. A major remanié horizon in the Speeton Clay – the base of bed D2D. The bed is marked by a horizon of corroded belemnites and scattered phosphatic nodules, the latter including internal moulds of ammonites. The photograph shows a well-preserved *Polyptychites* and (just beneath the clip of the biro) the alveolar view of a belemnite. The biro is 15cm long. *(Lower)*. Ammonites from the Speeton Clay. (left) A phosphatised *Prodichotomites complanatus* (Koenen) from the remanié horizon at the base of bed D2D: (right) *Aegocrioceras spathi* Rawson, an uncoiling ammonite (adjacent whorls not in contact) from bed C7A.

ITINERARY X 93

some of the more distinctive marker bands (e.g. the nodular beds LB4C, LB3D and LB3D) can be picked out. This part of the sequence is not very fossiliferous though the belemnites *Praeoxyteuthis, Aulacoteuthis* and *Oxyteuthis* occur; heteromorph ammonites are mainly flattened except in bed LB1A where whorl fragments of *Paracrioceras elegans* occur. Bed LB1 is highly pyritic and attractive small crystals occur in clusters.

2A. Speeton Cliffs. At Speeton Beck a footpath leads back to the holiday camp, though it is easier to return along the shore. Alternatively, one can continue southeastward to examine the Hunstanton (formerly Red Chalk) and Ferriby Formation Chalks beneath Speeton Cliffs. This second part of the itinerary can also be treated as a separate day, in which case it is possible to park near the church at Speeton village and follow the footpath from the church to the cliff top, then head down the undercliff to Speeton Beck (Figure 29). However, the shore route from Reighton, though slightly longer, is easier going.

To the southeast of Speeton Beck the Speeton Formation is buried beneath landslipped chalk and boulder clay and has rarely been seen on the shore. Isolated small patches of the top B and A beds are seen occasionally at the cliff foot. Black pyritic clays opposite Queen Rock yield Upper Barremian and Lower Aptian ammonites and the bivalve *Grammatodon securis*. Further along, the higher shore is strewn with chalk boulders which makes the walking more difficult; between them patches of red and grey calcareous clay (A Beds) are sometimes uncovered. Eventually the landslip area gives way to sheer chalk cliffs about a kilometre from Speeton Beck, and here the Upper Albian Hunstanton Formation is exposed in a downfaulted (or slipped) mass. The impure chalks and thin marls here are thicker than the typical condensed limestone facies inland and mark a gradation towards the equivalent but generally more argillaceous Rødby Formation of the offshore area. The belemnite *Neohibolites minimus* is common, with occasional brachiopods.

2B. Red Cliff Hole (TA 165751) is a recess where the Hunstanton Formation appears again on intertidal reefs and sometimes at the very foot of the cliff. The lower part of the Ferriby Formation (formerly Lower Chalk) is well exposed above. The lowest unit consists of the "Grey Band", about 3.7 m of greenish grey chalk with large pyrite crystals, which could be of latest Albian age. Above are about 5 m of pink to reddish Cenomanian chalk which can be mistaken for the deeper red Hunstanton Formation, especially when the latter is covered with beach shingle. Note that the colour change between the "Grey Band" and the overlying pink chalk is sharp but cuts across the bedding irregularly. Above the pink chalk is a series of rubbly-bedded grey and pinkish chalks with two thicker pink bands about 7.5 and 20.5 metres above the base. The chalks here are in a facies resembling griotte – anastomosing thin seams of marl enclosing nodules of chalk formed during an early stage of diagenesis (Jeans, 1980). Fossils are quite common though sometimes crushed – *Holaster, Pycnodonte, Aucellina* and brachiopods predominating.

Higher in the cliff, and inaccessible at this point, the Plenus Marl at the base of the overlying Welton Formation is visible; it lies about 44 m above the base of the Ferriby Formation.

The visitor should not walk beyond Red Hole as the sections are dangerous and the tide reaches the cliff foot in places. It is equally dangerous to scramble up the grassy slopes to reach higher beds.

ITINERARY XI

Thornwick Bay and North Landing, Flamborough

P. F. Rawson and F. Whitham

O.S. 1:25,000 Sheet TA 26/27 Flamborough or
1:50,000 Landranger 101 Scarborough
G.S. 1:63,360 Sheet 55 Flamborough

The localities described in itineraries XI to XIII lie in the area of the Flamborough Headland Heritage Coast. When approaching sections visitors are asked to keep to the marked paths. Further information on the Heritage Coast is available at the Heritage Centre at South Landing ravine (Itinerary XIII).

In the Flamborough area the coastline is deeply eroded from Thornwick Bay to Flamborough Head and there are several small bays which contain magnificent arches, caves and sea stacks cut into hard chalk – so hard that it has been used as a building stone and can be seen in some of the older buildings in the area, including the 17th century lighthouse. At low tide some of the caves and arches provide access to adjacent coves, from which there is no escape when the tide turns. Users of this guide are warned that **it is highly dangerous to stray beyond the confines of the bays described here.**

Throughout the Flamborough area the Chalk is overlain by a thick blanket of boulder clay, and it is the contrast between the two that causes such a marked change in slope half to three quarters of the way up the cliffs. In places, downwash from the clay smears the chalk while fallen lumps are soon broken up by the sea to release the enclosed erratic rocks and fossils. Hence, although the local beach shingle is predominantly of local flint and chalk it contains numerous exotic pebbles, including small carnelians. Spectacularly large boulders of various types, including Shap Granite, are scattered over the beaches.

ITINERARY XI

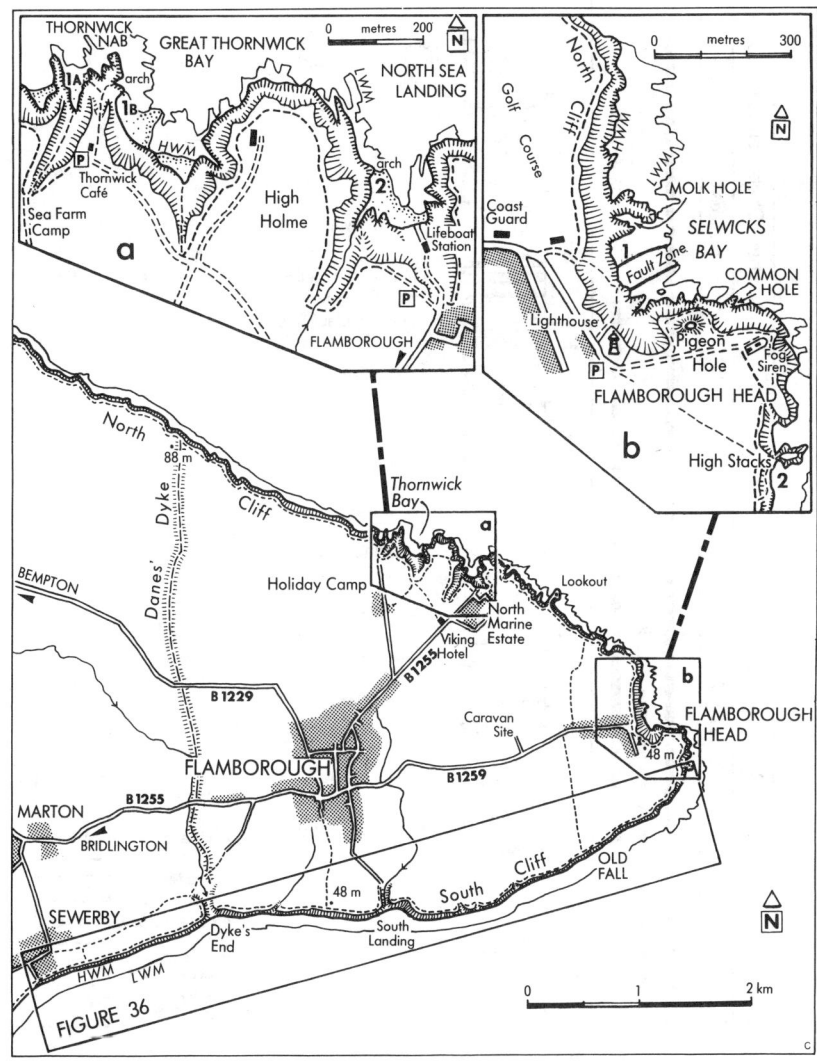

Figure 31. Locality map of the Flamborough area, with details of Thornwick Bay and North Landing. The rectangle extending from Sewerby to Flamborough Head is as in Figure 35.

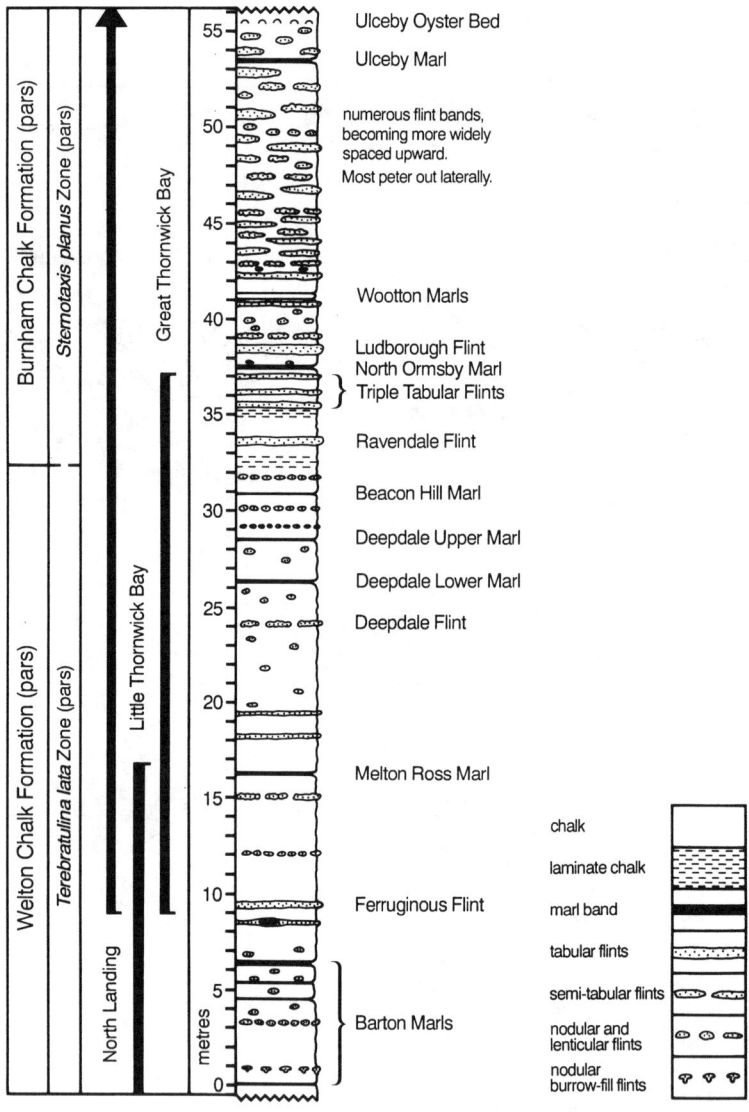

Figure 32. The Chalk sequence at Thornwick Bay and North Landing.

ITINERARY XI   97

Locality 1. Thornwick Bay (Welton and Burnham Formations).

From Flamborough village take the B1255 towards North Landing, turning immediately before "The Viking Hotel" (Figure 31) onto a track signposted "Thornwick Bay" for almost a kilometre until the track terminates at a parking area opposite the Thornwick Cafe. With care a coach can drive here. A footpath in front of the cafe leads to Thornwick Nab; the path soon forks, the left fork leading into Little Thornwick Bay and the right fork into Great Thornwick Bay.

In both bays the chalk is very sparsely fossiliferous and belongs almost wholly to the *Terebratulina lata* Zone of the Turonian. The exposures are of interest primarily to show how this single, ill-defined zone of the Chalk can be subdivided lithologically. The sequence embraces the upper half of the Welton Chalk Formation and the base of the overlying Burnham Formation, and contains a number of named marl and flint bands (Figure 32). The sediments accumulated along the southern margin of the Cleveland Basin, within the Howardian Flamborough Fault Belt, and are almost 10 metres thicker than in correlative sections on the northern part of the East Midlands Shelf.

1A. Little Thornwick Bay. The lowest beds are visible on the north side of the bay (Figure 33), where four deeply eroded, narrow clefts near low water mark (the lowest at the foot of the arch on the seaward side) represent the individual marl bands of the Barton Marls. The Ferruginous Flint about 3 metres higher is a prominent 15 cm thick tabular, carious flint with red-weathering patches. This bed can be traced on both sides of the bay and round Thornwick Nab into Great Thornwick Bay, forming one of the best marker bands. *Inoceramus lamarcki* occurs 2.75 m below this band. A second good marker is a 2-4 cm thick marl, the Melton Ross Marl, which forms a prominent line in the cliff at the head of Little Thornwick Bay.

1B. Great Thornwick Bay. From Little Thornwick return up the short cliff path and cross into Great Thornwick Bay. On either side of the path as it reaches the beach the highest flint band seen is the band of semi-tabular flints 5.4 m above the Ferruginous Flint. The latter is readily visible round the corner to the left, where it rises from near shore level to form the roof of the arch at Thornwick Nab (Figure 33). From here the succession can be traced across the scars to the southern and eastern sides of the bay. In the cliff on the southern side the lowest flint band is a prominent grey tabular flint, the Deepdale Flint, which rises westward from the foot of the cliff. Above are 2 deeply weathered notches formed by the Deepdale Lower and Upper Marls, and the sequence can be traced upward to the Ravendale and Triple Tabular flint bands. These form equally clear markers at the head of the bay on the eastern cliff. The suggested boundary of the *Terebratulina lata* and *Sternotaxis plana* zones is at the base of the 50 cm unit of thinly-bedded chalks directly below the Ravendale Flint, where the first *S. plana* occur. *Gibbithyris semiglobosa* is present just above the flint.

Once the Ravendale and Triple Tabular flints have been identified it is easy to follow the succession downwards along the eastern cliff to the Melton Ross Marl, which forms a deep cleft in the cliff foot at low water mark.

Locality 2. North Landing (Welton and Burnham Formations).

From Thornwick Bay either walk along the cliff top path eastwards or drive back to the B1255 and turn left to North Landing (Figure 31). The road terminates in a large car and coach park. From there walk down the cliff road adjacent to the cafe and pub. At the head of the bay on the left hand (western) side is a slight embayment and cave in the Chalk. A prominent ledge rising eastwards from beach level marks the position of a useful marker bed, the Ulceby Marl. A second marker horizon can be picked up about halfway along the western side of the Landing, at a conspicuous marine arch. Here the Ravendale Flint is about 2 metres above shore level, and just above it are the Triple Tabular Flint bands, with the prominent Ludborough Flint about half way up the inside walls of the arch. At the northwestern extremity of the Landing the remains of a ship's boilers lie in a deep cleft; a prominent rusty-brown flint band adjacent to them is the Ferruginous Flint. These three marker levels allow the whole succession (Figure 32) to be followed.

On the eastern side of the Landing the Beacon Hill Marl lies just above low water. A little higher in the sequence the thickest (20 cm) tabular flint, the Ludborough Flint, cuts across the mouth of Robin Lythes Hole, which leads into a magnificent cavern with another exit onto East Scar. The Ulceby Marl again forms a ledge, which descends to the shore in the cliff immediately adjacent to the lifeboat slipway.

Fossils are not common through most of the succession, but *Sternotaxis plana* occurs in the upper beds (from just below the Ravendale Flint). About 2 m above the Ulceby Marl is the Ulceby Oyster Bed, here about a 20 cm thick band of chalk with scattered oysters *(Pycnodonte vesicularis)* and occasional brachiopods.

Both this and the following itinerary occupy less than a full day but cannot be combined safely because by the time one is finished the tide will normally have risen too much for the other. Itinerary XIV provides an interesting "filler" to complete the day.

Figure 33 *(Upper)*. Little Thornwick Bay: the Barton Marls occur at the foot of the cliff where the waves are breaking, the Ferruginous Flint (FF) is 3 metres above, while the Melton Ross Marl (MR) forms a notch higher in the cliff. *(Lower)*. Thornwick Nab: the Ferruginous Flint (FF) is the prominent band half-way up the cliff.

ITINERARY XII

Flamborough Head

P. F. Rawson and F. Whitham

O.S. 1:25,000 Sheet TA 26/27 Flamborough or
1:50,000 Landranger 101 Scarborough
G.S. 1:63,360 Sheet 55 Flamborough

Locality 1. Selwicks Bay (structures in Chalk).

From Flamborough village follow the B1259 to Flamborough Head, where there is a large car park adjacent to the lighthouse (Figure 31). A path to the left of the lighthouse gate leads down the cliff to Selwicks Bay. A variety of lime-loving flowers, including orchids, can be seen on the boulder clay slopes adjacent to the path.

The main feature of geological interest is a zone of disturbance running E-W through the middle of the bay which represents part of the Howardian-Flamborough Fault Belt (Figure 34). The main zone of disturbance has a width of about 50 m, lying behind and to either side of the steps, within which the Chalk is folded, faulted, sheared and brecciated. There is no single fault plane, but where a small stream reaches the shore there is a spectacular change in dip either side of a minor fault close to the northern margin of the zone. In general the northerly dip decreases northwards. In the intertidal zone the small scars of chalk dip noticeably more steeply within the disturbance zone. The buttress of chalk forming the southern margin is strongly brecciated with calcite in vughs and interlocking veins.

There is some displacement of chalk across the zone, for to the south is flinty chalk of the top of the Burnham Chalk Formation and to the north flintless chalk of the basal Flamborough Chalk Formation. Detailed measurement of sections either side by F. Whitham coupled with measurements in Common Hole just east of the bay by Lamplugh (1895) suggest a minimum displacement of at least 21.5 metres. Movement along the zone appears to have embraced both extensional and compressional phases, and probably resulted from early Tertiary reactivation of deep-seated faults.

Away from the zone of disturbance the chalk on the north side shows numerous low angle (c. 45°) faults. It is often difficult to prove any vertical displacement but it is apparently only a few centimetres – sometimes in a normal direction, other times reversed. Horizontal slickensides also occur. Minor folds occur at several levels, especially about halfway up the cliff

Figure 34 *(Upper)*. The disturbed zone ("shatter belt") in Selwicks Bay: note the undisturbed chalk on the extreme left and right hand sides of the photograph. *(Lower)*. Disturbed chalk on the northern side of the contorted zone abutting against undisturbed chalk: a fault runs down the gully.

(see photograph on back cover), mainly developed above décollement horizons formed by thin marls.

Towards the northeast corner of the bay is a deep re-entrant, Molk Hole, leading to several small caves and two spectacular arches. At the foot of the cliff just before this re-entrant occurs the highest flint band (High Stacks Flint) of the Yorkshire Chalk, marking the top of the Burnham Formation. **This location should only be approached on a falling tide.**

To the south of the disturbed zone flinty chalk of the Burnham Formation shows far less low angle faulting than on the northern side. The small embayments here represent collapsed blow holes (note the concave chalk faces). The solitary sea stack is locally called "Adam"; its former partner ("Eve") on the opposite side of the bay was illustrated by Lamplugh (1896, plate 31), but has since been eroded away.

The flinty chalk in Selwicks Bay is referred to the lower *Hagenowia rostrata* Zone and the flintless chalk to the upper part of the zone. Fossils are uncommon but occasional examples of *Gonioteuthis westfalica*, *Echinocorys* sp., *Porosphaera globularis*, small brachiopods and fragmented inoceramids are found.

Locality 2. High Stacks (Flamborough Formation).

Return up the steps and walk round the lighthouse to a track leading eastwards to the foghorn (Figure 31). Follow the cliff-top path southeastwards to TA 257704, where a cliff path leads to the shore at High Stacks, a clay- and gravel-capped small promontory of chalk. From here about 167 m of chalk, forming all but the highest part of the Flamborough Formation, are exposed in continuous succession dipping gently westwards to Sewerby Steps, a distance of about 6.5 km (Figure 35). There are few distinctive lithological markers but some of the more prominent, thicker marls can be used to locate zonal boundaries (Whitham, in press). The cliff section from High Stacks to the next access point at South Landing ravine (about 3.4 km westwards) exposes the lowest part of the formation (lower part of the upper *Hagenowia rostrata* Zone), showing about 26 m of massive, very hard white chalk with a series of thin marls which increase in frequency up the succession to form a thinner-bedded sequence in the cliff near South Landing.

On the shore behind High Stacks the top flint of the Burnham Chalk is visible, while a deep cleft 1.5 m above the foot of the stack is an etched out 2 cm thick marl lying about 3.5 m above the base of the flintless Flamborough Formation chalk. The bulk of the chalk displays little lithological variation. It dips westwards from High Stacks for approximately

500 metres near to Old Fall, before levelling-out almost horizontally for about 2.5 km to within 400 metres of South Landing, where the low angle dip recommences to bring the highest beds in the cliff, including a 1-3 cm marker marl, near to shore level close to the ravine.

Fossils are fairly scarce in the main part of the section, but are more common at one level in a 1.5 m bed occupying the lower part of the cliff for over 2 km. This bed lies almost equidistant between High Stacks and South Landing and contains the tiny echinoid *Hagenowia blackmorei* (formerly misidentified as the larger index species, *H. rostrata*, which is confined to the lower *rostrata* Zone), *Gonioteuthis westfalica, Orbirhynchia pisiformis, Porosphaera globularis*, corals and echinoid spines. In the remainder of the chalk there are spasmodic occurrences of *Echinocorys* sp., fragmented inoceramids and sponges, the last occurring more frequently towards the end of the section. Large ammonites (? *Parapuzosia* sp.) have been recorded from South Landing. *H. blackmorei* is confined to about 4 m of accessible chalk some 22 m above the base of the formation on the east side of South Landing ravine and also occurs on the west side where a further 3.5 to 4 m of higher beds containing this echinoid are brought down to shore level by the westerly dip (see Itinerary XIII).

It is possible **(on a falling tide only)** to proceed along the shore from High Stacks to South Landing, returning via Flamborough. However, as the section is practically along the strike it is best, after reaching Old Fall, to return to High Stacks and Flamborough Head, noting on the way the large erratic boulders scattered over this part of the shore, including a Shap Granite about 1 m across.

ITINERARY XIII

South Landing to Sewerby

F. Whitham

O.S. 1:25,000 Sheet TA 26/27 Flamborough or
1:50,000 Landranger 101 Scarborough
G.S. 1:63,360 Sheet 65 Bridlington

Starting at Flamborough village take a minor road, signposted "South Landing", from the crossroads at TA 228702 (Figure 31). From the Heritage Centre at the car park a path leads down a ravine to the shore (Figure 35). On both sides of the Landing drift deposits mark a glacial melt-water channel cut through the chalk; the ravine follows this channel. A few years ago a storm cleared the beach in line with the ravine to reveal a chalk sequence distorted

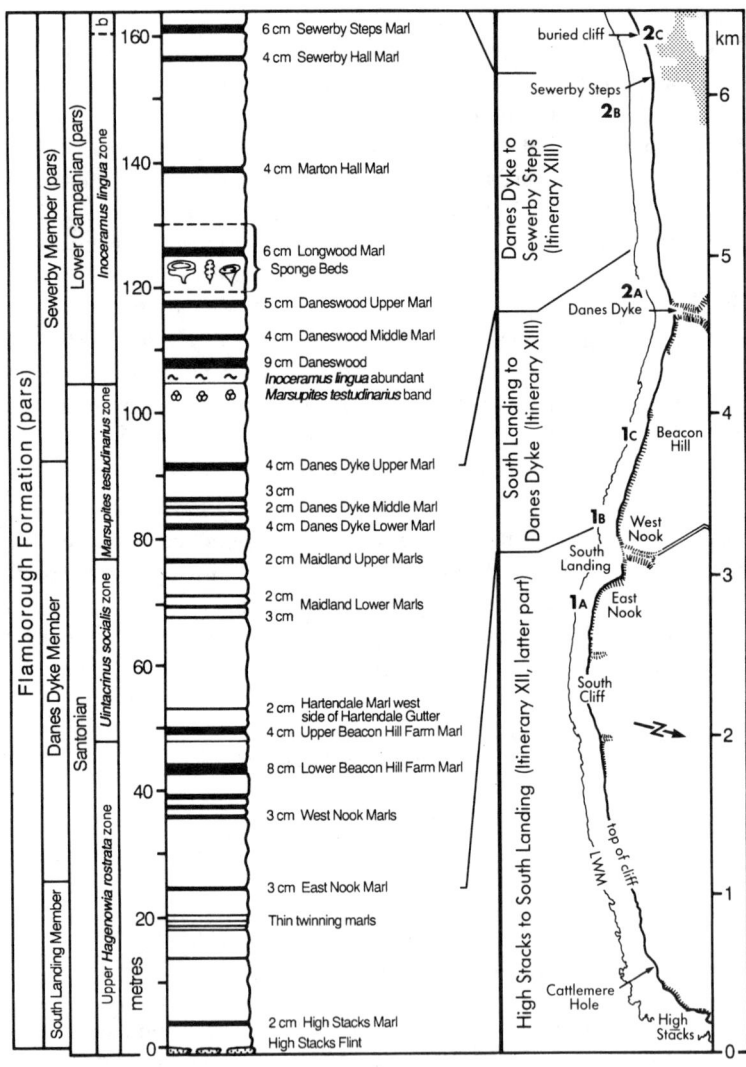

Figure 35. The Chalk sequence from High Stacks to Sewerby Steps.

by folding or faulting, probably reflecting another deep-seated structure in the Howardian-Flamborough Fault Belt. This faulted ground may have formed a line of weakness exploited by the glacial melt-waters.

Locality 1. South Landing to Dane's Dyke (Flamborough Formation).

1A. East Nook. To the east (left) of the Landing are exposed the highest beds described in Itinerary XII, which may be examined if not seen on the previous walk. After about 400 m the same sequence of beds as those observed at Old Fall appear. Return to the Landing.

The walk from South Landing westward to Sewerby Steps (TA 202686) is approximately 3 km and exposes about 138 m of the Flamborough Formation sequence, from the higher part of the upper *Hagenowia rostrata* Zone (c. 23 m) through the whole of the *Uintacrinus socialis* (29 m) and *Marsupites testudinarius* (26 m) Zones to the *Inoceramus lingua* Zone (just over 60 m exposed). Halfway along the section is the seaward end of Danes Dyke which is the only other exit from the beach before Sewerby. **Care must be taken as high tide reaches the base of the cliffs in a number of places.**

1B. West Nook. In the foot of the cliffs on the SW side of the Landing the bed containing *Hagenowia blackmorei* can be traced for about 100 m before dipping below beach level and provides a link with the last locality of Itinerary XII. This species appears to be confined to beds within an 8 m sequence spanning the ravine and is extremely rare outside this horizon. Only one or two occurrences are recorded elsewhere in the northern chalk. The remainder of the fauna in this part of the flintless rostrata chalk is restricted to occasional brachiopods, fragmented inoceramid shells, thick shelled *Echinocorys* sp., *Gonioteuthis westfalica granulata*, *Actinocamax verus*, and sponges (including *Amphithelion*, small varieties of *Laosciadia plana*, *Siphonia koenigi*, *Stichophyma tumidum* and abundant *Porosphaera globularis*).

1C. Beacon Hill. About 500 m from South Landing the boundary between the *H. rostrata* and *Uintacrinus socialis* Zones is marked by the first appearance of the zonal index in a bed of chalk approximately 2 m below an 8 cm thick marl (Lower Beacon Hill Marl) in the vicinity of Beacon Hill. The lower part of the *socialis* Zone comprises hard and soft chalks with thinner bedding and a few stylolitic horizons, while the higher part reverts back to more massive bedding. Stylolites are complex zig-zag contacts formed by loading and solution processes. A number of minor faults dissect the succession and several rock falls have taken place recently, the most massive one occurring about 200 m east of Hartendale Gutter (a sewage discharge outlet), obscuring a large part of the cliff.

In the past it has been said that this part of the coastal sequence is one of the very few areas in England to yield complete cups of *Uintacrinus socialis*. It is now almost impossible to find complete specimens in the lower, accessible parts of the cliffs, but isolated plates are common and at some horizons small groups of plates occur possibly all forming part of the same individual. Other fossils include fragmented inoceramids, *Echinocorys* sp., *Orbirhynchia pisiformis*, *Acutostrea boucheroni*, *Parasmilia* sp. and sponges belonging to the same group as listed for the previous zone. Belemnites of the *Gonioteuthis granulata* lineage appear to be less common and the occurrence of *socialis* plates diminishes in the higher beds, being absent in the highest 3 m of the zone.

Rare isolated plates of the zonal species *Marsupites testudinarius* first occur above the upper of two 2 cm marls (Upper Maidlands Marls) separated by several thin marls spread over about 2 m of chalk approximately 220 m from Danes Dyke. This horizon marks the base of the *testudinarius* Zone, which reaches its maximum thickness in this area. Massive bedding continues upward from the higher part of the preceding zone but the chalk becomes softer. The lowest 15 m of the zone occur in the cliff up to Danes Dyke, but *Marsupites* plates and other fossils are very rare here.

It is possible to return up the cliff at Dane's Dyke, following a footpath inland to Flamborough village.

Locality 2. Dane's Dyke to Sewerby Steps (Flamborough Formation).

2A. Dane's Dyke cuts into *testudinarius* Zone chalks. The seaward end forms part of an interglacial meltwater channel partly filled with drift which was re-excavated during the Bronze Age to form part of an entrenchment cutting across Flamborough Head. Faulting has been observed in the Chalk here when storms have cleared part of the beach (R. Myerscough, personal communication). The distance from here to Sewerby is about 1.5 km and the southwesterly dip exposes the remaining 11 m of *Marsupites* chalk and just over 60 m of *Inoceramus lingua* Zone chalk of which the highest 3 m above Sewerby Steps is assigned to the *Discoscaphites binodosus* Subzone.

Scars on the beach at the SW side of the dyke contain abundant *Marsupites* plates. In the cliff a 4 cm marl (Upper Danes Dyke Marl) can be correlated with the eastern side and marks a lithological change from more massive chalks to thinner bedded sequences parted by many thin marls. About 100 m along the section *Marsupites* plates become extremely common and complete calyces (cups) are found. This flood of the zonal species is spread over about 5 m of chalk in the middle part of the zone and it then dies out close to the upper boundary about 200 m SW of the dyke. Note the stylolitic horizons along this section.

Other species to be found in the remaining part of the zone include *Acutostrea boucheroni* (in bands), *Orbirhynchia pisiformis*, occasional specimens of *Echinocorys* sp., varieties of *Ventriculites, Porosphaera globularis* and other sponges often preserved as oxide films. Large examples of *Gonioteuthis granulata* occur.

About 2 m above the last appearance of *Marsupites*, and near to four seaweed-covered calcrete blocks (first noted by Rowe, 1904), the base of the *Inoceramus lingua* Zone is marked by a profusion of fragmented shells of the zonal species. This horizon also marks the boundary of the Santonian and Campanian stages. Chalk of the *lingua* Zone is for the most part fairly hard with some massive bedding interspersed with a series of thinly bedded horizons. Stylolitic surfaces are less frequent than in the previous zone. Fossils are common at some horizons with *Inoceramus lingua* the dominant bivalve while rare examples of *Sphenoceramus pinniformis* occur. Sponges are far more common in the *lingua* Zone than elsewhere, with the best developed concentration of hexactinellid and lithistid sponges occurring in the famous Flamborough Sponge Beds, which consist of just over 10 m of chalk, the basal beds lying some 15.5 m above the base of the zone. The more shallow dip of the strata where the Sponge Beds reach the shore provides a continuous exposure on the beach scars for a considerable distance, commencing about 350 m from Danes Dyke; the main exposure lies nearer to this ravine than to Sewerby.

Many fine sponges occur in both cliff and scars, including *Pachinion scriptum, Stichophyma tumidum*, varieties of *Laosciadia plana, Siphonia koenegi, Rhizopoterion cribosum, Amphithelion (Verruculina), Wollemannia laevis, Sporadoscinia strips, Leiostracosia punctata* and *Porosphaera globularis*. Also occurring in the Sponge Beds are very large *Echinocorys* (up to 80 mm long), *Inoceramus lingua* and *Sphenoceramus pinniformis*, with rare *Gonioteuthis granulata*. The top of the Sponge Beds is marked by three thinly bedded 20 cm chalk horizons spread over 1.5 m, with the intervening beds containing abundant *Acutostrea boucheroni*.

2B. Sewerby Steps. Above the Sponge Beds, towards Sewerby Steps, the more massive bedded chalk becomes less fossiliferous, with sporadic occurrences of *Echinocorys*, sponges, fragmented inoceramids and shell debris. Other species recorded in the *lingua* Zone here include a band of *Offaster pilula, Hagenowia* sp. large and rounded forms of *Cardiotaxis, Hypoxytoma tenuicostata, Orbirhynchia* sp. and the rare ammonites *Hauericeras pseudogardeni* and *Scaphites* sp. Echinoid spines and asteroid plates are common in the lower half of the zone. *Discoscaphites binodosus* occurs in the highest 3 m of chalk, above the steps.

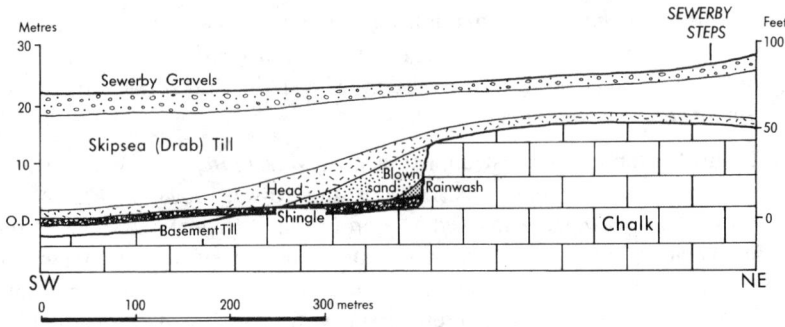

Figure 36. Diagrammatic section of the buried cliff at Sewerby (modified from Catt & Penny, 1966, pl. 24).

2C. The Buried Cliff (by P. F. Rawson). About 300 m SSW of Sewerby Steps is the buried cliff section described in the GA Guide to Hull (Penny, in Bisat et al., 1962, p. 18). Here, the modern cliff face shows the Chalk terminating abruptly against sands, shingle and boulder clay (Figure 36). The interface marks a Pleistocene chalk cliff which runs slightly obliquely to the modern cliff and can be traced for at least 50 metres before disappearing completely behind the glacial deposits. The buried cliff then strikes inland to run along the dip slope of the Chalk Wolds to the Humber (where it is visible at Hessle) and on into Lincolnshire. It is an interesting and important feature in the glacial chronology of the area (Catt & Penny, 1966). At the foot of the buried cliff is an interglacial shingle beach ("Sewerby Raised Beach"), about 1 metre above the modern beach level and resting on a planed surface of Basement Till, though the contact is only seen on the rare occasions when storms have stripped off the modern beach. The raised beach contains vertebrate remains indicative of the Last (Ipswichian) Interglacial, including the straight-tusked elephant *(Palaeoloxodon antiquus)*, the narrow-nosed rhinoceras *(Dicerorhinus hemitoechus)* and *Hippopotamus*. It dates to about 116,000 – 128,000 years ago.

The periglacial deposits (blown sand and head) above the raised beach mark the encroaching cold period of the Devensian (last) glaciation. As the climate deteriorated further ice spread over the area some 18,000 years ago to deposit the Skipsea Till, which blanketed the whole sequence. Over most of Holderness the Skipsea Till rests directly on Basement Till and the Sewerby succession is thus crucial in demonstrating that the tills represent deposition during two different glacial periods. The Basement Till probably represents the Wolstonian (penultimate) glaciation, about 140,000 years ago (Catt, 1990).

ITINERARY XIII 109

If the whole of the South Landing to Sewerby section is followed on a falling tide there will still not be time to return along the shore. Instead either go into Sewerby village and catch a bus back to Flamborough, or return along a footpath which follows the cliff top to Dane's Dyke and then strikes inland to Flamborough (Figure 32).

ITINERARY XIV

Langtoft, Foxholes and Staxton Hill

P. F. Rawson

O.S. 1:25,000 Sheet TA 06/16 Langtoft, 07/17 Hunmanby,
1:50,000 Landranger 101 Scarborough
G.S. 1:50,000 Scarborough, 1:63,360 Great Driffield

This brief itinerary links three localities along the B1249, which runs northwards across the Wolds from Driffield to Staxton. Two show inland exposures of chalk shatter zones in disused quarries immediately by the roadside, in which it is possible to park. The third, Staxton Hill, is an excellent viewpoint from which the glacial history of the Vale of Pickering area can be demonstrated. Combined with Itinerary XI or XII this can make a full field day, in which case drive from Flamborough to the west side of Bridlington and turn off the A165 onto the B1253 through Rudston. The churchyard at Rudston village contains the Rudston monolith, reputed to be the tallest standing stone in England. Dating to the Bronze Age, it is made of Jurassic sandstone and must have been transported for at least 16 km, possibly from the Pickering area (Allison, 1976). The grave of the East Riding's best-known novelist, Winifred Holtby, lies near the SW corner of the churchyard.

Continue westwards from Rudston to a traffic island at Octon Cross Roads where the B1253 crosses the B1249. Turn left (southwards) along the latter and drive through Langtoft; just south of the village on the east side of the road (TA 012659) is a disused chalk quarry in the Flamborough Chalk Formation (probably the *rostrata* Zone). The chalk at the southern end of the section is almost horizontal, but northwards it is dragged up to dip of about 50°, before passing into a zone of breceiated, calcite-veined and slickensided chalk beneath the grassy slopes at the northern end of the quarry. Note that in a quarry across the dale from here the chalk is undisturbed. This shatter zone probably reflects reactivation of the E-W trending Langtoft Fault, recently traced beneath the Chalk eastwards along the northern side of Bridlington and into the offshore area (Kirby & Swallow, 1987).

From Langtoft head northwards to Foxholes. About half a kilometre north of the village on the east side of the road (TA 012735) is another old quarry which exposes flinty chalk of the Burnham Formation. The main face shows a mass of chalk dipping more or less uniformly north at about 70°, whereas in the top right-hand corner of the quarry, above the grassed-over talus slopes, the chalk is almost horizontal. The contact between dipping and horizontal chalks is now difficult to see, but horizontal slickensides are sometimes visible. Vein calcite crystals can be picked up on the slopes beneath. This shatter zone again appears to lie over a pre-Chalk fault and extends eastwards to merge with the Bempton shatter zone on the coast.

Continue northward to the top of Staxton Hill and park at the picnic area and viewpoint (signposted). There are public toilets here, and tables in the picnic area. On a clear day there is a spectacular view of virtually the whole Vale of Pickering. In the foreground the scarp edge is formed of chalk underlain by the Hunstanton Formation and Speeton Clay, while the Vale is floored by Kimmeridge Clay. The dip slope of the Tabular Hills rises away from the observer in the distance (Figure 37). But the main reason for stopping here is to consider the glacial history and imagine the time when one would have been looking out from snow-capped hills over an extensive glacial lake. *The Geologists' Association* visited here on 3 August 1967, when Dr L. F. Penny gave a succinct account of the glacial evolution of the area (Penny & Rawson, 1969, p. 204); we are grateful to Dr Penny for allowing us to reproduce it in slightly modified form here:

"The Wykeham moraine, which curved half-way across the Vale of Pickering, marked the limit of an ice lobe which had entered the valley from the east. The other end of the valley was simultaneously blocked by Vale of York ice at the Ampleforth moraine. Lake Pickering extended between the two, depositing lake clays which have been proved in numerous boreholes. The lake was fed principally by the waters of the Newtondale spillway which deposited the delta on which Pickering stands; and also by those of the Forge Valley spillway which, hemmed in between the ice of the Wykeham lobe and the Corallian dip-slope, were forced westward, depositing the Hutton Buscel kame terrace and the delta fan which spread south-westward from the point where it entered the lake.

As the ice retreated from the Wykeham moraine, the waters of the Forge Valley were able to flow straight into the lake, initiating the present course of the Derwent and destroying the southern half of the Wykeham moraine (which probably abutted on the Chalk scarp around Ganton). Retreating still

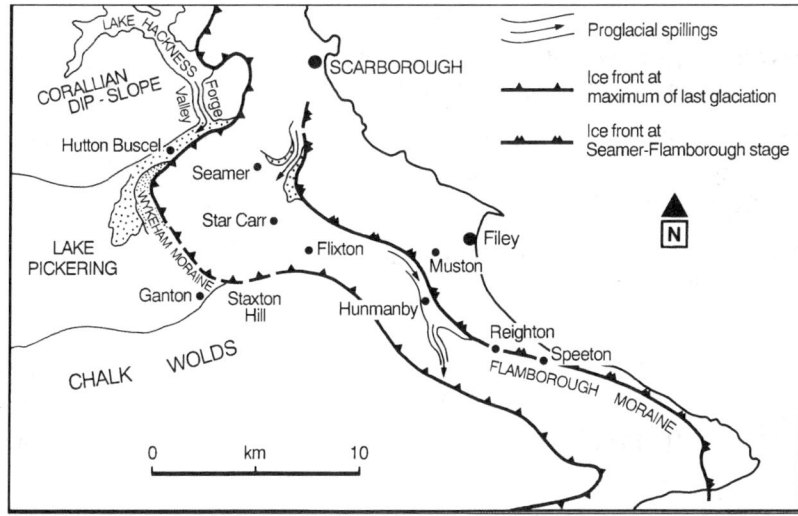

Figure 37. Glaciology of the eastern end of the Vale of Pickering (redrawn from Penny & Rawson, 1969, fig 3, by permission of the authors and the Geologists' Association).

farther, the ice uncovered the Seamer-Scarborough Valley, whose waters then similarly flowed into the lake and deposited the Seamer delta. The position of the ice front farther south-east at this time is uncertain, but it is probably related to the Flamborough moraine, for an ice margin drawn in this way encloses an area of markedly fresher glacial topography, which includes the hummocky ground between Muston and Hunmanby and the well-developed drumlins near Reighton.

All these events relate to a period at, and immediately after, the maximum of the Last Glaciation, but "Lake Pickering" probably remained in a diminished form and at a lower level until the Kirkham spillway had been finally cut down to its present level. Certainly there was a lake here in early post-Glacial times, the shores of which were inhabited by Mesolithic man around Flixton and Starr Carr. The valley was still fen in historical times, of which the many 'carrs' (fens) and 'ings' (water meadows) bear witness throughout the area, and is still liable to severe flooding, despite the cutting of the Hertford River and the canalisation of the Derwent in the eighteenth and nineteenth centuries."

REFERENCES

AGAR, R. 1960. Post-glacial erosion of the north Yorkshire coast from the Tees estuary to the Humber. *Proc. Yorks. geol. Soc.*, **32**, 409-428.

ALEXANDER, J. 1986. Idealised flow models to predict alluvial sandstone body distribution in the Middle Jurassic Yorkshire Basin. *Marine Petrol. Geol.*, **3**, 298-305.

ALLISON, K. J. 1976. *The East Riding of Yorkshire Landscape.* Hodder & Stoughton, 272 pp.

BAIRSTOW, L. F. 1969. Lower Lias. In (Hemingway, J. E., Wright, J. K. & Torrens, H. S., eds) *International Field Symposium on the British Jurassic. Excursion Guide 3, N. E. Yorkshire.* University of Keele, 47 pp.

BATE, R. H. 1959. The Yons Nab Beds of the Middle Jurassic of the Yorkshire Coast. *Proc. Yorks. geol. Soc.*, **32**, 153-164.

BISAT, W. S., PENNY, L. F. & NEALE, J. W. 1962. *Geology around the University Towns: Hull.* Geol. Ass. Guide No. 11, 34 pp.

BLACK, M. 1928. 'Washouts' in the Estuarine Series of Yorkshire. *Geol. Mag.*, **65**, 301-307.

BLACK, M. 1929. Drifted plant beds of the Upper Estuarine Series of Yorkshire. *Q. Jl geol. Soc. Lond.*, **85**, 389-437.

BLACK, M., HEMINGWAY, J. E. & WILSON, V. 1934. Summer field meeting in N. E. Yorkshire: report by the directors. *Proc. Geol. Ass.*, **45**, 291-306.

BUCKMAN, S. S. 1915. A palaeontological classification of the Jurassic rocks of the Whitby district, with a zonal table of Liassic ammonites. Pp 59-102 in Fox-Strangways & Barrow, q.v.

CATT, J. A. 1990. Geology and Relief. In (Ellis, S. & Crowther, D. R. eds). *Humber perspectives.* Hull University Press, 13-28.

CATT, J. A. & PENNY, L. F. 1966. The Pleistocene deposits of Holderness, East Yorkshire. *Proc. Yorks. geol. Soc.*, **35**, 375-420.

DEAN, W. T. 1954. Notes on part of the Upper Lias succession at Blea Wyke, Yorkshire. *Proc. Yorks. geol. Soc.*, **29**, 161-179.

DELAIR, J. B. & SARGEANT, W. A. S. 1985. History and bibliography of the study of fossil vertebrate footprints in the British Isles: Supplement 1973-83. *Palaeogeogr. Palaeoclimatol., Palaeoecol.*, **19**, 123-160.

REFERENCES

FARROW, G. E. 1966. Bathymetric zonation of Jurassic trace fossils from the coast of Yorkshire, England. *Palaeogeog., Palaeoclimatol., Palaeoecol.*, **2**, 103-151.

FOX-STRANGWAYS, C. 1892. The Jurassic rocks of Great Britain, vol. 1, Yorkshire. *Mem. geol. Surv. G.B.*, 1x + 551 pp.

FOX-STRANGWAYS, C. & BARROW, G. 1915. The Geology of the Country between Whitby and Scarborough. *Mem. geol. Surv. G.B.*, iv + 144 pp.

GODFREY, A. & LASSEY, P. J. 1974. *Shipwrecks of the Yorkshire Coast*. Dalesman Books, 168 pp.

GOWLAND, S. & RIDING, J. B. 1991. Stratigraphy, sedimentology and palaeontology of the Scarborough Formation (Middle Jurassic) of Hundale Point, North Yorkshire. *Proc. Yorks. geol. Soc.*, **48**, 375-392.

GREENSMITH, J. T., RAWSON, P. F. & SHALABY, S. E. 1980. An association of minor fining-upward cycles and aligned gutter marks in the Middle Lias (Lower Jurassic) of the Yorkshire Coast. *Proc. Yorks. geol. Soc.*, **42**, 525-538.

HEMINGWAY, J. E. 1963. Pp. 1-23 in *Geology of the Yorkshire Coast*. Geol. Assoc. Guide No. 34, 34 pp. (revised 1968).

HEMINGWAY, J. E. 1974. Jurassic. In (Rayner, D. H. & Hemingway, J. E. eds). *The Geology and Mineral Resources of Yorkshire*. Leeds (Yorkshire Geological Society). 161-223.

HEMINGWAY, J. E. & RIDDLER, G. P. 1982. Basin inversion in North Yorkshire. *Trans. Instn Min. Metall.* (section B), 91, B175-B186.

HEMINGWAY, J. E., WILSON, V. & WRIGHT, C. W. 1963. *Geology of the Yorkshire Coast*. Geol. Assoc. Guide No. 34, 34 pp. (revised 1968).

HOWARD, A. S. 1985. Lithostratigraphy of the Staithes Sandstone and Cleveland Ironstone Formations (Lower Jurassic) of north-east Yorkshire. *Proc. Yorks. geol. Soc.*, **45**, 261-275.

HOWARTH, M. K. 1955. Domerian of the Yorkshire Coast. *Proc. Yorks. geol. Soc.*, **30**, 147-175.

HOWARTH, M. K. 1962. The Jet Rock Series and the Alum Shale Series of the Yorkshire Coast. *Proc. Yorks. geol. Soc.*, **33**, 381-422.

HOWARTH, M. K. 1973. The stratigraphy and ammonite fauna of the Upper Liassic Grey Shales of the Yorkshire Coast. *Bull. Br. Mus. (Nat. Hist.)*, **24**, 235-277.

REFERENCES

HUDLESTON, W. H. 1878. The Yorkshire Oolites, Pt 2, the Middle Oolites. Section 2, the Coralline Oolites, Coral Rag and Supra-Coralline Beds. *Proc. Geol. Ass.*, **5**, 407-494.

JACKSON, J. W. 1911. A new species of *Unio* from the Yorkshire Estuarine Series. *The Naturalist, 1911*, 211-214.

JEANS, C. V. 1980. Early submarine lithification in the Red Chalk and Lower Chalk of Eastern England: a bacterial control model and its implications. *Proc. Yorks. geol. Soc.*, **43**, 81-157.

KANTOROWICZ, J. D. 1990. Lateral and vertical variation in pedogenesis and other early diagenetic phenomena, Middle Jurassic Ravenscar Group, Yorkshire. *Proc. Yorks. geol. Soc.*, **48**, 61-74.

KENDALL, P. F. 1902. A system of glacier lakes in the Cleveland Hills. *Q. Jl geol. Soc. Lond.*, **58**, 471-571.

KENT, P. E. 1980a. Subsidence and uplift in East Yorkshire and Lincolnshire: a double inversion. *Proc. Yorks. geol. Soc.*, **42**, 505-524.

KENT, P. E. 1980b. *Eastern England from the Tees to the Wash*. British Regional Geology, HMSO, vii + 155 pp.

KIRBY, G. A. & SWALLOW, P. W. 1987. Tectonism and sedimentation in the Flamborough Head region of north-east England. *Proc. Yorks. geol. Soc.*, **46**, 301-309.

KNOX, R. W. O'B. 1973. The Eller Beck Bed (Bajocian) of the Ravenscar Group of north-east Yorkshire. *Geol. Mag.*, **110**, 511-534.

KNOX, R. W. O'B. 1984. Lithostratigraphy and depositional history of the late Toarcian sequence at Ravenscar, Yorkshire. *Proc. Yorks. geol. Soc.*, **45**, 99-108.

LAMPLUGH, G. W. 1889. On the subdivisions of the Speeton Clay. *Q. Jl geol. Soc. Lond.*, **45**, 575-618.

LAMPLUGH, G. W. 1896. Notes on the White Chalk of Yorkshire. Part III. The Geology of Flamborough Head, with notes on the Yorkshire Wolds. *Proc. Yorks. geol. Soc.*, **13**, 171-191.

LINTON, D. L. 1955. The problem of Tors. *Geogr. Jl*, **121**, 470-487.

LIVERA, S. E. & LEEDER, M. R. 1981. The Middle Jurassic Ravenscar Group ("Deltaic Series") of Yorkshire: recent sedimentological studies as demonstrated during a field meeting, 2-3 May 1980. *Proc. Geol. Ass.*, **92**, 241-250.

MIDDLEMISS, F. A. 1976. Lower Cretaceous Terebratulidina of Northern England and Germany and their geological background. *Geol. Jb.*, **A30**, 21-104.

MILSOM, J. & RAWSON, P. F. 1989. The Peak Trough – a major control on the geology of the North Yorkshire coast. *Geol. Mag.*, **126**, 699-705.

MORRIS, K. A. 1979. A classification of Jurassic marine shale sequences: an example from the Toarcian (Lower Jurassic) of Great Britain. *Palaeogeog., Palaeoclimatol., Palaeoecol.*, **26**, 117-126.

NAMI, M. 1976. An exhumed Jurassic meander belt from Yorkshire. *Geol. Mag.*, **113**, 47-52.

NAMI, M. & LEEDER, M. R. 1978. Changing channel morphologies and magnitude in the Scalby Formation (Middle Jurassic) of Yorkshire (England). In (Miall, A. D., Ed.) *Fluvial sedimentation*. Can. Soc. Petrol. geol. Mem. 5, 431-440.

OWEN, J. S. 1985. *Staithes and Port Mulgrave Ironstone*. The Cleveland Industrial Archaeol. Res. Rep. no. 4, 41 pp.

PAGE, K. N. 1989. A stratigraphic revision for the English Lower Callovian. *Proc. Geol. Ass.*, **100**, 363-382.

PALMER, J. 1956. Tor formation at the Bridestones in North-east Yorkshire, and its significance in relation to problems of valley-side development and regional glaciation. *Trans. Inst. Br. Geog.*, no. **22**, 55-72.

PARSONS, C. F. 1977. A statigraphic revision of the Scarborough Formation. *Proc. Yorks. geol. Soc.*, **41**, 203-222.

PARSONS, C. F. 1980. The Aalenian and Bajocian Stages. In: Cope, J. C. W. *et al. A correlation of the Jurassic rocks in the British Isles. Part Two: Middle and Upper Jurassic*. Geol. Soc. Lond., Spec. Rep. 15, pp. 3-21.

PENNY, L. F & RAWSON, P. F. 1969. Field meeting in East Yorkshire and North Lincolnshire. *Proc. Geol. Ass.*, **80**, 193-218.

PHILLIPS, J. 1829. *Illustrations of the Geology of Yorkshire, Part 1. – the Yorkshire Coast*. London, xvi + 192 pp. (2nd edit. 1835, 3rd edition 1875).

POWELL, J. H. 1984. Lithostratigraphical nomenclature of the Lias Group in the Yorkshire Basin. *Proc. Yorks. geol. Soc.*, **45**, 51-57.

PYE, K. & KRINSLEY, D. H. 1986. Microfabric, mineralogy and early diagenetic history of the Whitby Mudstone Formation (Toarcian), Cleveland Basin, U.K. *Geol. Mag.*, **123**, 191-203.

RASTALL, R. H. & HEMINGWAY, J. E. 1940. The Yorkshire Dogger, 1, the coastal region. *Geol. Mag.*, **77**, 177-197.

RAWSON, P. F., CURRY, D., DILLEY, F. C., HANCOCK, J. M., KENNEDY, W. J., NEALE, J. W., WOOD, C. J. & WORSSAM, B. C. 1978. *A correlation of Cretaceous rocks in the British Isles*. Geol. Soc. Lond., Spec. Rep. 9, 70pp.

RAWSON, P. F., GREENSMITH, J. T. & SHALABY, S. E. 1983. Coarsening-upward cycles in the uppermost Staithes and Cleveland Ironstone Formations (Lower Jurassic) of the Yorkshire coast, England. *Proc. Geol. Ass.*, **94**, 91-93.

RAWSON, P. F. & MUTTERLOSE, J. 1983. Stratigraphy of the Lower B and basal Cement Beds (Barremian) of the Speeton Clay, Yorkshire, England. *Proc. Geol. Ass.*, **94**, 133-146.

RAWSON, P. F. & RILEY, L. A. 1982. Latest Jurassic – Early Cretaceous events and the "Late Cimmerian Unconformity" in North Sea area. *Bull. Am. Ass. Petrol. Geol.*, **66**, 2628-2648.

RIDING, J. B. 1984. A palynological investigation of Toarcian to early Aalenian strata from the Blea Wyke area, Ravenscar, North Yorkshire. *Proc. Yorks. geol. Soc.*, **45**, 109-122.

RIDING, J. B. & WRIGHT, J. K. 1989. Palynostratigraphy of the Scalby Formation (Middle Jurassic) of the Cleveland Basin, north-east Yorkshire. *Proc. Yorks. geol. Soc.*, **47**, 349-354.

ROMANO, M. & WHYTE, M. A. 1987. A limulid trace fossil from the Scarborough Formation (Jurassic) of Yorkshire: – its occurrence, taxonomy and interpretation. *Proc. Yorks. geol. Soc.*, **46**, 85-95.

SARGEANT, W. A. S. 1970. Fossil footprints from the Middle Trias of Nottinghamshire and the Middle Jurassic of Yorkshire. *Mercian Geologist*, **3**, 269-282.

SELLWOOD, B. W. 1970. The relation of trace fossils to small-scale sedimentary cycles in the British Lias. *Geol. Jl Spec. Iss.* **3**, 489-504.

VERSEY, H. C. 1939. The Tertiary History of East Yorkshire. *Proc. Yorks. geol. Soc.*, **23**, 302-314.

WHITHAM, F. 1991. The stratigraphy of the Upper Cretaceous Ferriby, Welton and Burnham Formations north of the Humber, north-east England. *Proc. Yorks. geol. Soc.*, **48**, 227-254.

WHITHAM, F. in press. The stratigraphy of the Upper Cretaceous Flamborough Formation north of the Humber, north-east England. *Proc. Yorks. geol. Soc.*

WILSON, V. 1949. The lower Corallian rocks of the Yorkshire coast and Hackness Hills. *Proc. Geol. Ass.*, **60**, 235-271.

WOOD, C. J. & SMITH, D. 1978. Lithostratigraphical nomenclature of the Chalk in North Yorkshire, Humberside and Lincolnshire. *Proc. Yorks. geol. Soc.*, **42**, 263-287.

WRIGHT, J. K. 1968. The stratigraphy of the Callovian rocks between Newtondale and the Scarborough coast, Yorkshire. *Proc. Geol. Ass.*, **79**, 363-399.

WRIGHT, J. K. 1972. The stratigraphy of the Yorkshire Corallian. *Proc. Yorks. geol. Soc.*, **39**, 225-266.

WRIGHT, J. K. 1977. The Cornbrash Formation (Callovian) in North Yorkshire and Cleveland. *Proc. Yorks. geol. Soc.*, **41**, 325-346.

WRIGHT, J. K. 1978. The Callovian succession (excluding Cornbrash) in the western and northern parts of the Yorkshire Basin. *Proc. Geol. Ass.*, **89**, 239-261.

WRIGHT, J. K. 1983. The Lower Oxfordian (Upper Jurassic) of North Yorkshire. *Proc. Yorks. geol. Soc.*, **44**, 249-281.

YOUNG, G. & BIRD, J. 1822. *A Geological Survey of the Yorkshire Coast.* Clark, Whitby, iv + 322 pp. (2nd edit. 1828).

NOTES

NOTES

NOTES